Electronic Textiles

Electronic Textiles

Editor

Jamshed A Khan

scitus academics

Electronic Textiles

Edited by **Jamshed A Khan**

ISBN: 978-1-68117-261-3
Library of Congress Control Number: 2016934804

SCITUS Academics LLC,
www.scitusacademics.com
Box No. 4766, 616 Corporate Way,
Suite 2, Valley Cottage,
NY 10989

Printed in United States of America on Acid Free Paper

Preface

Electronic textiles, also known as smart garments, smart clothing, smart textiles, or smart fabrics, are fabrics that enable digital components, and electronics to be embedded in them. Smart textiles are fabrics that have been developed with new technologies that provide added value to the wearer. E-textiles are an exciting, new field of electronics that combines embedded electronics with fabric and textiles. These wearable electronics are gaining lots of attention and are becoming more accessible to the non-technical crowd by the day. Smart textiles can be broken into two different categories: aesthetic and performance enhancing. Aesthetic examples include everything from fabrics that light up to fabrics that can change color. Some of these fabrics gather energy from the environment by harnessing vibrations, sound or heat, reacting to this input. Then there are performance enhancing smart textiles, which will have a huge impact on the athletic, extreme sports and military industries. There are fabrics that help regulate body temperature, reduce wind resistance and control muscle vibration – all of which help improve athletic performance. In general, the development of electronic textiles supports the idea of wearable computing, or electronic devices worked into garment designs. However, there are other applications of electronic textiles, such as interior design technologies, that also rely on integrating electronic components into fabrics or fibers. Within the relatively new industry around electronic textiles, projects are pursued based on various different kinds of functionality. Some electronic textile products may only be built for data storage, while others provide physical interfaces through control elements in the garment design. Although the practical applications of this technology are very new, some projects involving electronic textiles and intelligent garments do exist now. Many of these are aimed at uses within the health care industry, where wearable computing elements can be used to monitor vital signs and otherwise track a person's health in real time. This book, Electronic Textiles, emphasizes on thenovelty and usages of smart garments or fabrics in numerous fields.

Table of Contents

CHAPTER 1

E-textiles in Clinical Rehabilitation: A Scoping Review

Amanda Fleury [1,2], **Maddy Sugar** [1] **and Tom Chau** [1,2,*]

[1] *Bloorview Research Institute, Holland Bloorview Kids Rehabilitation Hospital, Toronto, ON M4G-1R8, Canada*

[2] *Institute of Biomaterials and Biomedical Engineering, University of Toronto, Toronto, ON M5S-3G9, Canada*

ABSTRACT

Electronic textiles have potential for many practical uses in clinical rehabilitation. This scoping review appraises recent and emerging developments of textile-based sensors with applications to rehabilitation. Contributions published from 2009 to 2013 are appraised with a specific focus on the measured physiological or biomechanical phenomenon, current measurement practices, textile innovations, and their merits and limitations. While fabric-based signal quality and sensor integration have advanced considerably, overall system integration (including circuitry and power) has not been fully realized. Validation against clinical gold standards is inconsistent at best, and feasibility with clinical populations remains to be demonstrated. The overwhelming focus of research and development has been on remote sensing but the opportunity for textile-mediated feedback to the wearer remains unexplored. Recommendations for future research are provided.

KEYWORDS

electronic textiles; rehabilitation; healthcare; remote monitoring

1. INTRODUCTION

The term "e-textiles" (short for electronic textiles) is often used interchangeably with "smart fabrics", "functional apparel" and even "wearable technology". "Wearable technology" broadly refers to the category of electronics that are carried on the body, while "e-textiles" refers specifically to garments or textiles with some sort of technology, such as wiring and sensors, embedded directly into the fabric.

As early as 1998, researchers have been working on garments to monitor physiological signals for healthcare, military and space exploration [1]. In the pioneering Wearable Motherboard garment [1], rigid electronic components for sensing and processing were attached to a fabric substrate. This work catalyzed a vision in which clothing serves as a platform for computing. Over the last 10 years, e-textile technologies have evolved from hobby crafts and fashion innovations to healthcare solutions. In parallel, there has been a surge of related research, with a Scopus search yielding some 100 papers in 2014 alone with the key word "e-textile", compared to only 6 in 2002. This growth of e-textile research and clinical application can be attributed to a number of technological advances as detailed below.

1. Sewable and washable microcontrollers like the commercially available, Arduino Lilypad (lilypadarduino.org) offer researchers affordable and fabric-friendly embedded platforms for quickly developing robust prototypes [2].

2. Flexible circuits have allowed the textile integration of advanced electronics capable of sensing and information transmission, without compromising the comfort of the wearer. Likewise, smaller and flexible power sources for e-textile applications have emerged [3] while energy-harvesting power electronics are already under development by companies like PowerLeap.(www.powerleap.com).
3. Conductive threads and yarns, usually composed of stainless steel or conductive silver with a nylon core, have become widely available. The properties of such threads and fabric transmission lines for textile computing applications have been documented [4,5,6], facilitating their selection and implementation in e-textile solutions.
4. Fabric-based components including capacitors, resistors, and transistors have been devised [4,7,8,9], allowing full integration of simple circuits onto a fabric substrate. Piezoelectric textiles have also emerged, promoting the development of a full range of fabric-based sensors [10,11].

With a globally aging population, it is not surprising that the majority of applied research on e-textiles has focused primarily on the unobtrusive, remote monitoring of older persons [12,13], particularly post-myocardial infarction or post-stroke. Increasingly, these wearable monitoring systems are capable of tracking complex movement patterns [14]. Against this backdrop, this paper provides a timely and critical appraisal of recent developments in textile-based sensors with either demonstrated or potential applications in clinical rehabilitation. Unlike previous e-textile reviews (e.g., [12]), our appraisal is not limited to physiological or behavioral monitoring applications. Before presenting our review methodology and findings, we introduce an integration rating. These introductory remarks will help to contextualize the ensuing review.

1.1.Degree of Integration

Fundamental to the concept of an e-textile, is the degree to which electronic components are integrated into the fabric. In this paper, we consider a fully integrated sensor as one that is woven directly into the fabric, in the absence of any rigid components. Conversely, an unintegrated sensor is one that is independently packaged and simply attached to the user. Table 1 proposes a rating system to describe the degree of integration.

Table 1. Degree of integration.

Rating	Description
0→	Wearable computer: no textile integration
1→	Superficial integration: components in pockets or connected to fabric with snaps
2→	Partial integration: some sensing components incorporated (e.g., woven, knit, printed, embroidered *etc.*) into fabric
3→	Partial integration: all sensing components incorporated (e.g., woven, knit, printed, embroidered *etc.*) into fabric
4→	Partial integration: wiring and sensing woven into fabric
5→	Towards full integration: sensing, wiring and power supply woven into fabric

The underlying aim of this review is to explore literature relating to the application of e-textile technology. A scoping review will allow us to survey the existing literature in an inclusive way while facilitating the identification of gaps and topics for further research.

2. METHODS

We searched the literature for recent studies related to the use of e-textiles in rehabilitation. The search was conducted using electronic databases including IEEE, PubMed, Scopus, and ISI Web of Science. We used various combinations of the terms outlined in Table 2.

Table 2. Terms used in literature search.

Textile	Electronic	Rehab
fabric clothing	smart intelligent sensing	physiology biomechanics bio-

We included all English-language, peer-reviewed journal articles that directly examined the development or implementation of e-textile sensors and actuators with applications in healthcare and rehabilitation, in the last five years (2009–2013). With a specific interest in integrated biomechanical and physiological sensors, we excluded all chemical sensors, and sensors which primarily fall into the category of "wearable technology"; that is, lacking any integration.

3. RESULTS

Our search yielded 100 unique articles, 35 of which were found to meet the aforementioned criteria. Various characteristics of the studies are summarized in Table 3. We were specifically interested in the intended application within clinical rehabilitation, the stage of development (e.g., conceptual design, laboratory characterization without human participants, or empirical testing with human participants), sensor type (e.g., electrical, optical, inertial, thermal), the degree of integration (Table 1) and the status of sensor validation (e.g., from unvalidated to validation against a gold standard).

A variety of physiological and behavioral measurements, relevant to clinical rehabilitation, have been enabled through the integration of sensors within a fabric substrate. Some sensors, such as electrocardiogram (ECG) electrodes, are commonly used and the degree of integration is such that an acceptable quality signal is possible from what looks and feels like simple fabric. Other sensors are in their infancy, little more than proof-of-concept, and are therefore often as bulky and obtrusive as their standard non-textile counterparts. The results of our search are presented below by the phenomenon measured. Each of the ensuing sections follows a common format: introduction to the phenomenon of interest, current measurement practice, textile innovations, and, merits and limitations of the current textile technology. The review closes with some recommendations for future research. Figure 1 summarizes the number of studies appraised by phenomenon (horizontal axis) and degree of integration (shading).

Table 3. Summary of recently published e-textile research (DoI = degree of integration; BP = blood pressure; SpO_2 = saturation of peripheral oxygen; VOP = venous occlusion plethysmography).

Reference	Stage of Development	# Study Participants	Participant Type	Sensed phenomenon	Sensor validation	DoI	Area Application of
Adnane *et al.* [15]	Empirical testing	1	Target	ECG, respiration	Non-fabric alternative: 3-lead ECG; pneumography	2	Sleep disorders
Angelidis [16]	Conceptual design	N/A	N/A	ECG, BP, SpO_2, temperature, sweat	N/A	5	Healthcare: general
Baek *et al.* [17]	Empirical testing	5	Adult, male, healthy	ECG, pulse, BP	Non-fabric alternative: Biopac ECG& PPG, Finometerpro blood pressure	2	Hospital monitoring, Remote monitoring
Bianchi *et al.* [18]	Empirical testing	24	Adult, healthy	ECG	Gold standard; clinical polysomnography	4	Sleep disorders
	Empirical testing	50	Target, database	ECG			
Cho *et al.* [19]	Empirical testing	2	Adult, male, healthy	ECG	None	4	Remote monitoring
Fletcher *et al.* [20]	Lab testing	N/A	N/A	EDA, pulse	Gold standard	4	Emotion
	Empirical testing	12	Adult, healthy	EDA, acceleration, tempreature	Gold standard		
Gioberto and Dunne [21]	Lab testing	N/A	N/A	Strain	None	3	Monitoring: general
Giorgino *et al.* [22]	Empirical testing	3	Unknown	Strain	Human expert	4	Motor rehabilitation
Goy *et al.* [23]	Lab testing	N/A	N/A	VOP	N/A	2	Remote monitoring
	Empirical testing	5	Adult, healthy	VOP	Gold standard: 4 Ag/AgCl electrodes		
Hannikainen *et al.* [24]	Empirical testing	9	Adult, healthy; Target	Bioimpedance	None	3	Remote monitoring

Table 3. *Cont.*

Reference	Stage of Development	# Study Participants	Participant Type	Sensed phenomenon	Sensor validation	DoI	Area Application of
Harms *et al.* [25]	Computer model	5	Adult, healthy	Acceleration	None	N/A	Motor rehabilitation
Hong *et al.* [26]	Empirical testing	18	Adult, healthy	ECG	Non-fabric alternative: 3 lead ECG	4	Remote monitoring
Kim and Cho [27]	Empirical testing	12	Target	BP, HR	N/A	2	Treatment
Lanata *et al.* [28]	Lab testing	N/A	N/A	EDA	Platinum electrodes	N/A	Emotion detection
	Empirical testing	35	Adult, healthy	EDA	Ag/AgCl electrode	4	
Lee *et al.* [29]	Empirical testing	Unknown	Unknown	ECG, respiration, pulse wave velocity, EMG, pressure	Non-fabric alternatives: commercial sensors	3	Remote monitoring
Lee *et al.* [30]	Empirical testing	15	Adult, male, healthy	Knee joint movements via bioimpedance	Non-fabric alternative: tilt sensor	3	Motor rehabilitation
Lee *et al.* [31]	Empirical testing	1	Adult, healthy	EDA, pulse wave	None	4	Remote monitoring
Lee and Chung [32]	Empirical testing	1	Adult, healthy	ECG, acceleration	None	2	Remote monitoring
Li *et al.* [33]	Lab testing (mathematical modeling)	N/A	N/A	Temperature	Mathematical model	2	Remote monitoring; diagnostic tool
Lofhede *et al.* [34]	Empirical testing	5	Adult, healthy	EEG	Standard EEG electrodes	3	Neonatal monitoring
Lopez *et al.* [35]	Lab testing	N/A	N/A	ECG, temperature, acceleration, position	Simulated signals	2	Hospital monitoring
	Empirical testing	5	Target	ECG temperature, acceleration, position	None		
Lorussi *et al.* [36]	Lab testing	N/A	N/A	Bend angle	Non-fabric alternative: electrogoniometer	0	Motor rehabilitation
	Empirical testing	1					

Table 3. *Cont.*

Reference	Stage of Development	# Study Participants	Participant Type	Sensed phenomenon	Sensor validation	DoI	Area Application of
Marquez *et al.* [37]	Empirical testing	3	Adult, male, healthy	Bioimpedance	Gold standard: clinical bioimpedance spectrometer	3	Remote monitoring
Preece *et al.* [38]	Empirical testing	20	Adult, healthy	Strain	Non-fabric alternative: AMTI force platforms	4	Motor rehabilitation
Di Renzo *et al.* [39]	Lab testing	N/A	N/A	Posture	Non-fabric alternative: "traditional" ECG (no further details provided)	N/A	General e-textiles
	Empirical testing		Target	ECG, respiration		4	
Schwarz *et al.* [40]	Lab testing	N/A	N/A	Electroconductivity	Mathematical model	N/A	General e-textile
Shu *et al.* [41]	Empirical testing	8	Adult, male, healthy & Target	Pressure	Gold standard: force platform; commercial in-sole pressure system	4	Hospital monitoring
Song *et al.* [42]	Empirical testing	3	Adult, healthy	ECG	None	3	Healthcare: general
Tormene *et al.* [43]	Empirical testing	1	Adult, male, healthy	Strain	Non-fabric alternative: triaxial accelerometer & magnetometer	4	Motor rehabilitation
Vuorela *et al.* [44]	Empirical testing	1	Adult, healthy	ECG, respiration	Gold standard: pneumotachograph; clinical ECG (# leads not specified)	2	Remote monitoring
Yamada *et al.* [45]	Lab & empirical testing	1	N/A	Strain	None	1	Motor rehabilitation
Zhang *et al.* [46]	Lab testing	N/A	N/A	ECG, respiration, SpO_2	Signal database, lung simulator & patient simulator	2	Remote monitoring
	Empirical testing	15	Adult, male, healthy	Respiration	Gold standard: clinical ventilator tester		
	Empirical testing	10	Adult, male, healthy	ECG	Non-fabric alternative: Polar HR monitor		
Zheng *et al.* [47]	Lab testing	N/A	N/A	Power	Conventional discharge policies	1	General e-textiles
Zysset *et al.* [48]	Empirical testing	N/A	N/A	Temperature, acceleration	None	3	General e-textiles

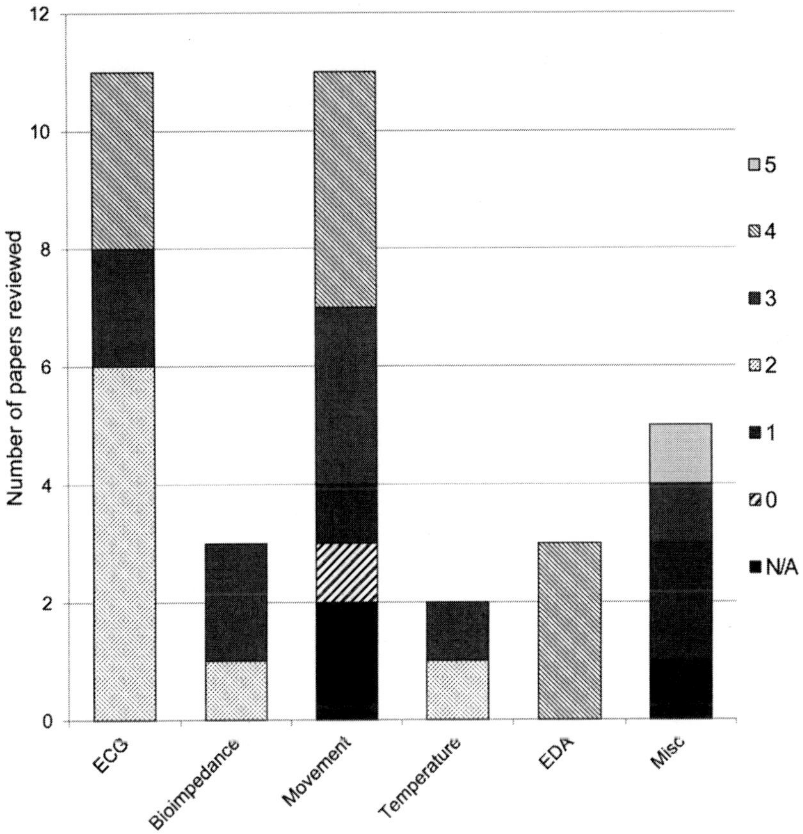

Figure 1. Summary of the reviewed studies by phenomenon measured (horizontal axis) and degree of textile integration (shading of the stacked bars).

3.1. Electrocardiogram (ECG)

3.1.1. Phenomenological Background

The heart continuously circulates blood throughout blood vessels of the entire cardiovascular system by means of systematic muscular contractions. A specialized group of pacemaker cells known as the sinoatrial node (SA node) spontaneously generates rhythmic electrical impulses that initiate myocyte contraction [49]. Heart rate (HR) is determined by the firing rate of the SA node, which spreads the electrical signal in the form of cell membrane depolarization events throughout the atrium. Upon reaching the atrioventricular node (AV node), the depolarization wave then propagates to the ventricles. The contraction of the heart muscle creates a pressure gradient that ejects blood throughout the body.

The SA node is innervated by both sympathetic and parasympathetic nervous system fibers and coordinates efferent responses that meet the body's variable metabolic demands and mediate responses to physiological stress. Stress-induced sympathetic stimulation is known to increase the firing rate of electrical signals in the SA node whereas "rest and digest" parasympathetic activity is known to decrease SA node activity. Conditions where heart rate is too fast, too slow, or irregular are known as arrhythmias. Heart rate variability (HRV), specifically, refers to beat-to-beat variations in the interval between heartbeats and may be used diagnostically in patients post-myocardial infarction [50].

3.1.2. Current Practice

Methods of heart rate measurement involve detection of cardiac pulse and frequency, as signified by periodic physiological changes. A tactile palpation of the pulse may be used clinically to determine HR, but many other technologies provide detailed recordings of various circulatory system characteristics.

In electrocardiography (ECG), electrical potential impulses initiated by the SA node are conducted by ionic currents and detected by electrodes on the skin's surface. Traditionally, an electrolyte solution is applied between the electrode and the skin to facilitate the conversion of ionic currents into electrical currents by oxidation-reduction reactions [51]. Each cardiac cycle reveals typical waves of depolarization and repolarization that are used to calculate the frequency of heart muscle contractions. Typically the normal-to-normal (NN) interval is used to measure heart rate, which refers to the time interval between successive sinus node depolarizations.

Electrodes are positioned on either side of the heart in pairs and conventionally require a conducting gel. Figure 2 depicts an electrical model of a wet electrode-skin interface [52]. In this model, Vh,Re and Ce are the half-cell potential of the electrode (dependent on type of material), its resistance and its capacitance. Rg is the effective resistance of the gel; Vs is the potential due to the semi-permeability of the stratum corneum, the top layer of the epidermis; Rs and Cs are the resistance and capacitance of the epidermal layer and Rd is the resistance of the deeper skin layers. While the number of electrodes may vary, 12-lead ECG monitoring is typically accepted as the "gold standard" in arrhythmia diagnosis [53]. Classic ECG monitoring requires that the patient remain stationary in the supine position and is thus not conducive to continuous ambulatory monitoring. Wearable Holter devices facilitate long-term and ambulatory monitoring of general wellness, sport, sleep, and numerous heart irregularities, often providing data for heart rate variability analyses [18,19,39].

The balancing actions of both sympathetic and parasympathetic branches of the ANS are represented in measurements of heart rate. Moreover, changes in heart rate may demonstrate arousal in the sympathetic or parasympathetic nervous system. It may also be possible for ECG to distinguish the contributions of both ANS branches by detecting different frequencies [54]. Heart rate variability (HRV), therefore, can be used as an indicator of specific physiological or emotional changes [55].

3.1.3. Textile Innovations

3.1.3.1. Sensor development

ECG is typically detected in a clinical setting using pre-gelled electrodes (Ag/AgCl) attached by adhesive to the skin. Textile electrodes, in contrast, are usually composed of conductive threads woven directly into fabric or by applying a metallic conductive coating to conventional fabrics. Both gelled and ungelled textile electrodes have been proposed.

Gel-based electrodes are said to have better signal quality [56,35], while gel-free electrodes have been touted as being more comfortable and less irritating over time [57].

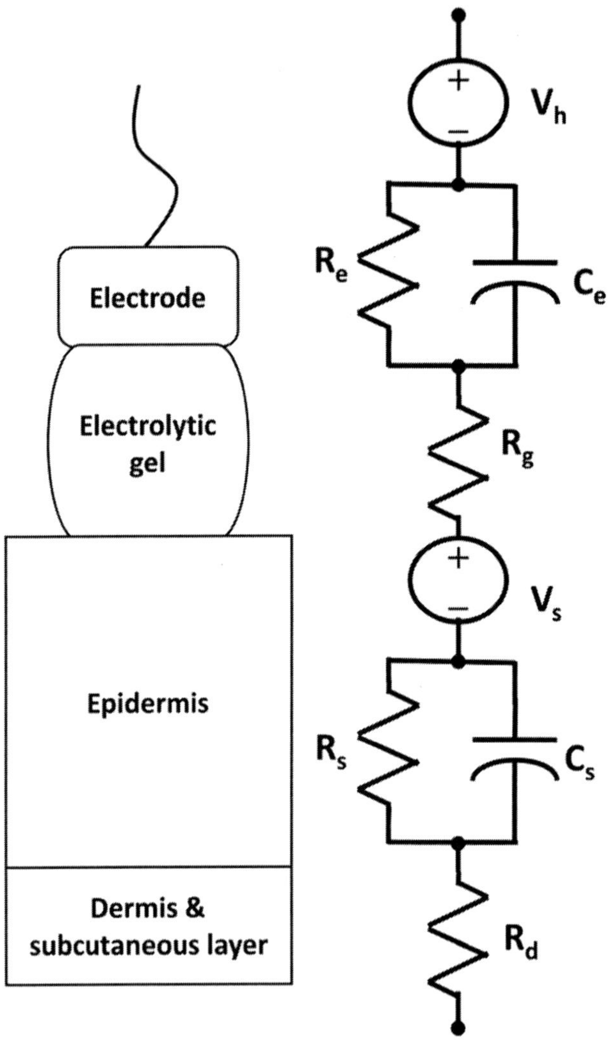

Figure 2. Electrical model of biopotential measurement at the skin surface.

Recent advancements in signal processing techniques, electrode specifications, and garment design have attempted to reduce the disparity between the two signals.

Improving fabric-based ECG signals

Lee and Chung [32] addressed common signal quality issues with textile electrodes during activity (walking and running) by using adaptive filtering techniques. Data from an on-board three-axis accelerometer were used to remove motion artifact, resulting in improved ECG signal quality, although observations were based primarily on visual inspection of the signals. Similarly, the accelerometer signals were used as a measure of physical activity, but the ability to distinguish between rest, walking, and running was not verified quantitatively. In a subsequent experiment, Hong et al. [26] presented a quantitative comparison of their single-lead fabric-based ECG with a conventional 3-lead system for treadmill walking/running at variable speeds (3 km/h, 6 km/h, and 9 km/h). In an experiment with 18 healthy adult male participants, the authors found high correlations between the two systems on all measures, including HRV (>0.96), R-peak detection, and wave duration, suggesting the suitability of fabric-based ECG for ambulatory monitoring.

PVDF

Lee et al. [29] turned to sensor design, exploring a different type of sensor for physiological measurement, namely polyvinylidene fluoride (PVDF) thin films, which change their electrical properties in response to applied physical stress. The authors investigated the use of PVDF and thermoplastic polyurethane (TPU) hybrids to measure respiration, ECG, pulse wave velocity, and muscle activity, among other signals. Qualitative results were presented for a number of experiments with different sensors, but ECG signals from PVDF sensors appeared similar to those obtained from standard Ag/AgCl electrodes. Baek et al. [17] also employed PVDF sensors to measure the ballistocardiogram (BCG; a measure of the ballistic force caused by the ejection of blood from the heart during the cardiac cycle) in an office chair designed for unconstrained health monitoring during daily activities.

Active electrodes

In addition to using redundant PVDF sensors, Baek et al. [17] sought to improve ECG signal acquisition by using active electrodes, which have high input impedance and perform signal amplification as close to the sensor as possible. The authors quantitatively compared the results from ECG measurements through a driven-capacitive-ground circuit (composed of two active electrodes embedded in the back rest of an office chair and a single fabric-based electrode in the seat) to those using a conventional ECG measurement system, reporting mean beat-to-beat errors of less than 3.0%. Similarly, Zhang et al. [46] proposed a shirt with three integrated active Ag/AgCl button electrodes for single-lead ECG monitoring. In an offline analysis with 10 adult male participants, ECG signals from the sensing shirt were compared with those from a commercially available HR monitor (Polar Electro, Finland). Good agreement between textile HR estimates and Polar reference measurements (mean difference of 0.24 beats per minute [46]) was achieved. Both Baek et al. [17] and Zhang et al. [46] reported accurate QRS detection using their textile systems.

Improving skin-electrode interface

Several advancements have also been made in garment design and fit to improve skin-electrode contact and hence signal quality, as well as to reduce motion artifact during ambulatory monitoring of ECG. Focusing specifically on the properties of jacquard woven electrodes, Song et al. [42] argued that woven textiles can be manufactured more easily and consistently than knits, and with more uniform properties. However, the jacquard woven electrodes were not quantitatively compared to either knits or standard Ag/AgCl electrodes. The authors compared different weaves and concluded that woven electrodes with a reduced number of conductive warp yarns and conductive paste between connections provide the best signal-to-noise ratio. In contrast, Cho et al. [19] investigated several different garment designs to decrease motion artifact of ECG electrodes in a knit shirt, relying on the convex shape of the embroidered electrodes and the proximity of the electrodes to the body as facilitated by the stretchable knit. Their different garment designs involved lines of less elastic fabric (8% polyurethane vs. 20% in the rest of the garment) in various configurations to stabilize the shirt. The authors concluded that the cross-type garment, with lines of stiffer fabric in a cross shape over the heart, offered the smallest displacement of the textile electrodes during prescribed movements and the best signal to noise ratio during a 5 min. walking task.

Sensing shirt/belt for sleep measurements

In several independent studies, wearable ECG has been investigated for use in sleep disorder diagnosis. The current standard in sleep disorder diagnosis and assessment is polysomnography (PSG), which requires multiple sensors (EEG, ECG, EMG, and respiration, among others) and expert interpretation. With an interest in home-based sleep monitoring, several groups have investigated the possibility of reducing the sleep parameters to heart rate variability alone [18,58], or ECG in combination with a respiration signal [15]. Using a previously validated smart shirt (www.smartex.it) equipped with e-textile ECG sensors, Bianchi et al. [18] investigated the possibility of assessing sleep quality using ECG alone. Using signals from a database, authors were able to detect apnea events with accuracies greater than 86% using several features of the ECG waveform. Although the implications for textile sensing were promising, comparisons between textile sensing and standard PSG were not disclosed. Seeking a simple method to detect apnea and hypopnea in at-risk patients, Adnane et al. [15] used a cardiorespiratory monitoring belt containing conductive fabric electrodes for ECG and PVDF thin films for respiration. The authors presented results from a single subject, demonstrating the ability of the system to detect apnea/hypopnea events using a ratio of the low and high frequency components of the respiration signal (similar to HRV).

Monitoring

There has also been interest in wearable technology and fabric-based sensors that provide patient comfort and ease of movement within a hospital-based environment. In this context, mobility is less important since patients generally move within a specific area, and health care professionals expect a higher signal quality than that currently attainable in mobile applications. The user interface is especially critical in a hospital environment to support workflow and minimize errors [35]. Lopez et al. [35] presented a system overview of an in-hospital patient monitoring system, termed LOBIN. The system included a wireless monitoring system, data acquisition and processing, and indoor location awareness for multiple patients. The authors reported unit tests of the individual components in a laboratory and systems integration tests in a hospital environment with cardiac patients. The system measured a variety of signals including ECG, respiration, and temperature, but only the electrodes for ECG were textile-based. Unfortunately, textile results were purely qualitative. The authors acknowledged that signal quality and robustness to motion were improved by the use of a conductive substance on the electrodes (water or gel), although these were prone to dry out over time.

3.1.4. Merits and Limitations

In commercial applications, fabric-based HR monitoring technology has been available in relatively inexpensive sports bras and chest belts from companies like Lululemon (www.lululemon.com) and Adidas (www.adidas.com), with a focus on personal feedback during training activities. E-textile ECG sensors may be able to fill the gap between commercial technologies (which often only display heart rate and possibly respiration rate to the user), and clinical systems (which are time consuming to apply and generally require an expert for placement of the electrodes and/or analysis of the results). Unfortunately many of the ECG systems mentioned here do away with feedback entirely and the results are intended to be sent directly to a clinician or health care professional for interpretation.

Rather than trying to replicate clinical testing or results, some researchers [18,15] have focused on deriving acceptable quality signals from body worn e-textile sensors for practical monitoring and diagnosis. By producing tools capable of screening and monitoring as opposed to a definitive diagnosis, researchers have been able to reduce the multiplicity of necessary sensors or electrodes, alleviating the burden on the user and perhaps minimizing the need for highly trained personnel for simple screening and monitoring applications.

Although applications in home or ambulatory monitoring generally do not require medical grade signals, textile measurements would be more meaningful upon contextualization against a gold standard. In many cases, textile ECG electrodes have only been assessed qualitatively, primarily via visual inspection of ECG waveforms [15,19,29,32,35].

3.2. Bioimpedance

3.2.1. Phenomenological Background

Bioimpedance is the opposition of current flow through biological tissue. In particular, cell membranes act as high impedance (i.e., capacitance) insulators at DC and low frequency currents. As a result, ions flow exclusively through extracellular space and thus the associated impedance is largely resistive [59]. Generally, the greater the cell concentration, the higher the low frequency impedance. High frequency currents on the other hand pass through cell membranes, and thus the ensuing impedance is a combination of capacitance and resistance, reflecting both intra- and extracellular conductivity [60]. Bioimpedance is thus frequency-dependent. In the simplest model of tissue impedance, ion-rich extracellular and intracellular spaces are represented as separate resistors while cell membranes are depicted as a capacitor, as shown in

Figure 3. The human body can be modeled as five cylindrical conductors (2 arms, 2 legs and trunk) [61] each with its own segmental impedance.

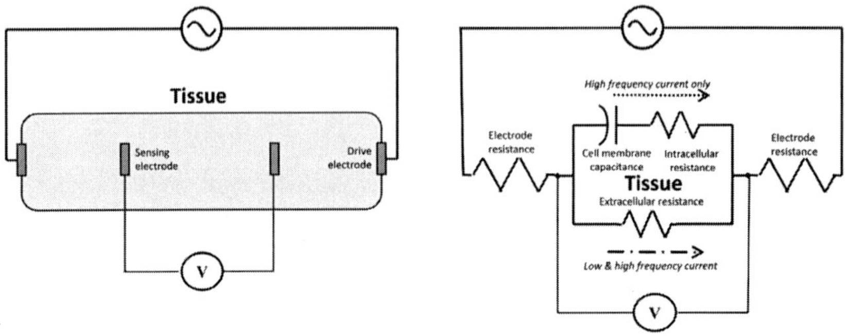

Figure 3. Tetrapolar measurement of tissue impedance (**left**); and equivalent electric circuit model (**right**).

3.2.2. Current Practice

The aforementioned biochemical currents in body tissue can be generated by or transformed into electrical currents in metallic conductors via electrodes and an accompanying electrolyte [62]. Given that its DC potential is relatively independent of DC current flow, the silver–silver chloride (Ag/AgCl) electrode is preferred in many clinical applications.

Conventional bioimpedance measurement deploys a tetrapolar arrangement of pre-gelled electrodes, where current is delivered through two drive electrodes and voltage is measured via two sensing electrodes (Figure 3). With this configuration, the impedances of the drive electrodes do not factor into the voltage measurement. The location of drive and sensing electrodes varies with different bioimpedance analysis (BIA) methods. For whole-body impedance, the drive and sensing electrodes are co-located on the hand and foot [61]. For segmental BIA (separately estimating appendicular and trunk bioimpedances), the current drive electrodes are situated on the right hand and foot, while the sensing electrodes are placed over the shoulders, clavicle, left elbow and wrist, and, right ankle and thigh [63]. Localized BIA, which aims to detect changes in soft tissue hydration and cell membrane integrity in a specific area of the body, situates electrodes over the segment of interest [64].

Bioimpedance measurements are typically made with an AC drive current. Single-frequency BIA typically uses a 50 kHz current. Multi-frequency BIA deploys a small set of frequencies while bioelectric spectroscopy sweeps through a wide range of frequencies [60].

The measured impedance values can be used to estimate total body water, intracellular and extracellular fluid volumes, fat free mass, and body cell mass [60]. These estimates can be useful for body composition analysis [65], assessing pulmonary function [66], and monitoring nutritional status of patients undergoing hemodialysis [67,68], among other applications.

3.2.3. Textile Innovations

Textile-based bioimpedance measures have focused on body composition [24,37] and segmental volume estimation [23,44]. Hannikainen et al. [24] proposed a total body water estimate using textile electrodes (material not disclosed), each connected by conductive yarn to a metallic snap fastener. The electrodes were sewn into elastic bands, which were worn around the user's wrist and ankle. With this tetrapolar configuration, the authors made multifrequency (5 to 50 kHz) impedance recordings during walking, cycling and running in nine participants. Although they did not validate their measurements against a clinical gold standard, the authors qualitatively demonstrated the ability to track changes in total body water levels during exercise conditions.

In a more rigorous quantitative study, Marquez et al. [37] estimated 8 body composition parameters in 3 male subjects using "textrodes" made from silver-plated, stretchable and highly conductive knitted fabric (78% polyamide, 22% elastomer; plated with 99% conductive silver) layered with a thin foam and an outer knit fabric to improve surface contact. Like Hannikainen et al., Marquez et al. also invoked a hand-foot tetrapolar arrangement, but swept through a broader range of frequencies (3 to 500 kHz) and validated their textrode measures against those of traditional Ag/AgCl electrodes. Encouragingly, Marquez et al. found that body composition estimates obtained from textrode-measured impedances were not statistically different than those derived via traditional metal electrodes.

As in body composition analysis, there have also been advances in impedance-based segmental volume estimation using textile electrodes. Vuorela et al. [44] created 225 mm2 square-shaped textile electrodes from conductive silver yarn. Deploying a tetrapolar configuration, they qualitatively compared bioimpedance-based estimates of thoracic volume to those obtained with a pneumotachograph, during quiet sitting, arm swinging and light exercise. The authors qualitatively ascertained a maximum difference of 50 mL in tidal volume estimates between the two methods. Turning to the lower extremity, Goy et al. [23] systematically investigated the utility of five different conductive textile materials (formed into 2 cm × 21 cm surface electrodes) for impedance-based estimation of blood volume changes, both in vitro (agar-agar synthetic skin) and in 5 healthy volunteers. Compared to standard Ag/AgCl electrodes, Goy et al. obtained the strongest signal via two-way stretch knit fabric made of silver coated nylon/elastomer thread (76% nylon, 24% elastic fibre; 100% Ag coating) and deemed textile electrodes as suitable for lower extremity venous occlusion plethysomography.

3.2.4. Merits and Limitations

Although textile electrodes have historically had signal quality issues due to poor skin contact resulting in an increased sensitivity to disturbances [44], recent advances in electrode characteristics and garment design have improved signal quality [37]. Similarly, wearability and data quality have advanced from contradictory bioimpedance measures using a jogging suit with bulky attachments for electronic components [24] to a streamlined system of textile electrode-bearing wrist and ankle straps, which yield impedance-based body composition parameters comparable in value to those derived via standard methods [37].

While it is possible to measure changes in total body water, fat-free body mass, or respiration rate with compelling accuracies (compared to clinical gold standards) using bio-impedance analysis, it is not possible to obtain absolute values. Further research is needed in particular with clinical populations (e.g., hemodialysis patients) to verify results obtained from healthy volunteers.

Finally, some lingering technical challenges seem apparent. Hannikainen et al. did not explain suspect findings for both cycling and walking conditions while Vuorela et al. vaguely alluded to data loss due to instrumentation “malfunction” during longer-term recordings. Goy et al. reported that to avoid errors due to instabilities of the polarization potential, the textile electrodes must rest on skin for 500 s prior to measurement, a constraint which may preclude practical clinical application.

3.3. Movement and Posture

3.3.1. Phenomological Background

Human movement is governed by the hierarchical control of the primary motor, premotor and supplemental motor cortices, basal ganglia, cerebellum, brainstem and spinal pattern generators, and modulated by feedback from the vestibular and sensory systems [69]. Likewise, postural maintenance integrates the control of higher brain centers (cerebellum and basal ganglia), brain stem and spinal cord with feedback from various sensory receptors to stabilize one’s center-of-mass via muscle contractions [70]. The measurement of functional movement and posture can be informative in rehabilitation science and practice. For example, such quantification can reveal changes in movement performance due to development, injury, recovery, therapy or assistive technology. The measurement of movement can also be used to decode communicative intent in non-verbal individuals [71] and to predict adverse events, such as falls or loss of balance [72]. Long-term compliance to therapy, particularly in the community, may benefit from remote monitoring through wearable systems that sense movement. Such portable systems enable therapists to track the frequency and quality of exercises performed outside of a clinic environment.

3.3.2. Current Practice

While expensive multi-camera-based systems that track body markers, such as the VICON system, are the gold standard for quantifying movement and body position (kinematics), their use is generally restricted to a finite volume within an institutional setting [73]. Mobile applications often invoke a combination of accelerometers, gyroscopes and goniometers to achieve the same purpose, but these sensors are unable to detect postural changes.

In similar vein, the kinetic characterization of movement has relied on floor-embedded force plates, which have a high degree of accuracy and precision, but are limited in size and flexibility. In-shoe pressure-sensing systems offer more flexible measurement possibilities but are still costly and unless wireless, may require tethering the footwear to wearable data acquisition units. With fabric-based sensing, there is an opportunity to explore alternative mobile monitoring of movement for motor rehabilitation and assessment.

3.3.3. Textile Innovations

While accelerometers and gyroscopes are appropriate for ambulatory measurement of limb and body movements, they are typically deployed as rigid sensors on the body or sewn into textiles, rather than fully integrated into fabric [74]. Early textile-based movement sensors used various conductive [73] or optical fibers [75,76] to estimate joint angles or limb position, but were not wearable for physical activity.

Recent advancements in textile-based measurement of movement have come primarily from the development of conductive elastomer (CE) sensors. Lorussi et al. [77] present a CE electrogoniometer, realized by applying two thin CE films (1 mm thick) to each side of a flexible and inextensible substrate. Their mathematical model indicated that the angle estimation error within the normal physiological range of bending is significantly reduced by the double layer configuration (<1% at a curvature of approximately 1 rad as opposed to approximately 3% with the single layer configuration). Although the substrate in this case was not fabric, Lorussi et al. [77] demonstrated that small, local deformations of the sensor do not significantly affect the overall angle measurement.

Lorussi et al. [36] subsequently employed the sensors on an extensible substrate—elastic woven cotton "Kinesio" tape capable of up to 40% elongation from its resting length. The final sensor consisted of three insulating layers of Kinesio tape and two CE sensing layers. Sensor readings were quantitatively compared to a traditional goniometer in lab tests as well as in a briefly described in vivo experiment measuring wrist flexion. In lab tests, predicted angles using CE sensors were generally within 1° of the imposed angle, as verified by the standard goniometer. Angles were not significantly different for small perturbations to local curvature.

Adopting a different manufacturing approach, Giorgino et al. [22] screen-printed 19 CE stretch sensors on a garment over the arm, shoulder, and chest areas to classify a specific set of upper limb postures (sagittal flexion, lateral abduction and external rotation of the shoulder) relevant to rehabilitation exercises. The study aimed at evaluating the contribution (measured by information gain [78]) of each sensor in order to inform optimization of sensor number and placement in later prototypes. Results predictably showed that sensors over locations with the most stretch during specific movements were most informative and that the information gain of each sensor depended on the exercise being performed. Nonetheless, information gain was consistent across the three subjects tested. Tormene et al. [43] later applied 13 similar sensors to the back of a fabric corset in superior-inferior and lateral directions to measure trunk motion in a single subject. While the sensors were not able to measure extension or lateral bending due to sensor buckling, for trunk flexion the fabric sensors correlated well with an accelerometer and magnetometer combination MEMS inertial sensor (Spearman's rank correlation coefficient = 0.88). Using a small training set of one example from each class (30, 60, 90 degrees of flexion), a strong level of agreement (Cohen's Kappa 0.85 ± 0.12) was reported between fabric-based and inertial sensor-derived estimates of flexion angles.

Yamada et al. [45] critiqued conductive elastomers and composites used to date as strain gauges, citing the slow response and creep as major concerns. Instead, the authors introduced a sensor composed of stretchable carbon nanotubes with desirable material properties including a fast response (14 ms), low overshoot (3%) and a fast recovery (5 s). The sensors maintained these properties when stretched to 150% for up to 10,000 cycles. While the sensors were insensitive to other types of deformation (twisting and applied force), the properties did change with temperature and the presence of certain gases, requiring that the sensors be sealed before use.

Shu et al. [41] employed similar strain sensors (knit fabric with a silicon coating doped with conductive carbon black) in an in-shoe plantar pressure measurement system. The strain sensors were sandwiched between two textured pieces of rubber, allowing an applied pressure from the foot to be measured as a change in length of the sensor. Six individual sensors were arranged in a shoe insole to measure pressures at the heel and metatarsals. In the 8 subjects tested, the sensors performed well in static measures of center of pressure (COP), with less then 5% measurement difference from the ANSI force platform used for comparison. With increased comfort, sensitivity, and fatigue resistance, this in-shoe measurement system may prove to be extremely useful for ambulatory and dynamic measurement outside of a clinic environment where expensive integrated force plates are impractical. Unfortunately, although the authors described a dynamic walking test, no quantitative results were reported. On a related front, Preece et al. [38] created an instrumented sock to identify heel strike, heel lift, and toe off during the gait cycle. The sock contained knitted strain sensors formed using the characteristics of the knitting pattern, which allowed loops of conductive yarn to come into or out of contact with each other as the fabric stretched, decreasing or increasing the effective length of the conductor and proportionally, the resistance. Although the sock solution bears clinical appeal, it was not designed for the measurement of absolute static or dynamic pressures. Nonetheless, it faired well at its intended purpose of event detection during the gait cycle.

Other fabric-based strain sensors have been implemented using conductive thread and specific machine stitches including cover stitch [79] and overlocked sensors [21]. In these sensors, the nature of the stitch allows loops of thread to come into or out of contact with each other as the fabric stretches, decreasing or increasing the effective length of the conductor and proportionally, the resistance. Using an overlock stitch, sensors produced a repeatable and approximately linear response up to 29% elongation. While these types of sensors are promising in their ease of application and repeatable response, the sensor properties, including baseline drift and recovery times, appear to be dominated by substrate characteristics.

Moving away from strain sensors, Lee et al. [30] investigated a system that measures joint motion by using bioimpedance to measure the volume changes in the muscle surrounding a joint, which can be associated with joint movement. Standard systems use disposable electrodes, but fabric-based electrodes have the advantage of being able to cover a larger surface area (1 cm × 25 cm in this case) and to be used for longer periods of time. The electrode in this case was polyester with Ni-Cu-Ni applied via electroless plating. The authors aimed to determine the optimum placement of fabric electrodes for discerning various joint angles, but did not compare their results to those with standard disposable Ag/AgCl electrodes. While the authors recommend the sensor pair most distal to the knee joint, their analysis is based on the ranking of each pair according to bioimpedance changes and signal-to-noise ratio (SNR), separately for hip and ankle movement. The three rank values are then added together to determine the best pairing. This type of analysis ignores the absolute values of the bioimpedance and SNR changes, which might suggest a more robust pairing.

Harms et al. [25] proposed a wrinkle modeling method for estimating the impact of loose-fitting garments on the automatic recognition of 10 shoulder rehabilitation exercise postures via a 'smart shirt' outfitted bilaterally with tri-axial accelerometers in its sleeves. The algorithm was shown to be a valuable tool for rapid prototyping, providing a prediction of the performance of textile-integrated, soft silicon-packaged electronics and thereby facilitating the optimization of smart garments before implementation. Di Rienzo et al. [39] developed an approach to quantify garment behavior during trunk motion with the aim of reducing motion artifact during high levels of activity. Their approach relied on a system of cameras and reflective markers (a total of 65 over the front and back of the torso) tracking changes in the relative positions of markers during movement with and without the garment. The final stage involved textile experts interpreting results to propose changes to fabric structure, electrode position, and garment cut.

3.3.4. Merits and Limitations

In the garments described by Giorgino et al. [22] and Tormene et al. [43], the sensing components and much of the wiring were integrated using a conductive elastomer. This produces a comfortable and fitted garment with only the battery and electronics as hard components, but CE sensors tend to exhibit creep, and languid response and recovery times. Sensors that yielded high sensitivity and repeatability, unfortunately have yet to be implemented on a fabric substrate [36,45,77].

In many cases, the effectiveness of posture recognition technology has been evaluated on its ability to correctly identify the posture of the participant, and not necessarily against standardized kinematic measurement tools [22,43,45]. As a result, no full garments have yet been described in the literature with the ability to discern joint angle with a high degree of accuracy. In addition, none of the studies described here test sensors with participants with disabilities; testing has occurred exclusively with healthy volunteers.

3.4. Temperature

3.4.1. Phenomenological Background

While heat generated by metabolic activity is distributed evenly throughout bodily tissues by circulating blood, gradients in temperature exist between core and periphery [80]. One mechanism of homeostatic thermoregulation involves microcirculatory adjustments that control conductive and convective heat loss between the skin and the external environment. Blood flow to the skin surface, where heat exchange takes place, can be adjusted through vasoconstriction and vasodilation of arteriovenous anastomoses (AVAs) [81]. Vasoconstriction of AVAs shunts blood toward, while vasodilation of AVAs directs blood away from, cutaneous vasculature. Therefore, the anatomy of the underlying cutaneous vasculature determines the degree to which heat exchange can take place.

AVAs are innervated by sympathetic nerve fibers, which receive and integrate thermoregulatory signals from the hypothalamus. The direction and magnitude of the sympathetic response is dependent on baseline body temperature, therefore it is possible that a certain stimulus can elicit either a vasodilation or vasoconstriction response [82]. Temperature variations among skin surface locations exist due to variations in circulatory anatomy, the degree of nervous system innervation and proximity to signaling organs [83]. Moreover, different skin surface locations respond more or less markedly to sympathetic stimulation. For instance, studies using infrared thermography of the face have shown that nasal and periorbital regions exhibit the most discriminatory skin temperature changes when sympathetically stimulated [84].

Transient thermoregulatory changes in skin blood flow occur in response to sympathetic stimuli while longer-term thermal homeostasis acts to maintain an individual's core body temperature within an optimal range. Deviation from this 'normal' temperature set point, such as in fever, reveals important diagnostic information [81].

3.4.2. Current Practice

The simplest measurements of temperature can be taken with technologies such as thermometers that equate thermal expansion in liquids and solids with numerical values, and require materials that are highly sensitive to heat changes. Resistance temperature detectors are highly sensitive as they exploit the repeatable, heat-induced rise in electrical resistance of a metal (e.g., copper) to sense temperature. A thermocouple is a rugged alternative that senses temperature as the voltage difference between a junction of two dissimilar metals at the probe tip and a reference junction (the same metals at a known temperature). A thermistor is typically a ceramic or polymer resistor whose resistance varies measurably and repeatedly with temperature. Generally, measurement range is greatest with thermocouples and most restricted with thermistors. A microbolometer consists of an array of uncooled thermal sensitive detectors (e.g., indium gallium arsenide) and are used in contemporary infrared thermal cameras.

In hospitals and clinics, thermometers are commonly used to collect oral, rectal, tympanic and axillary body temperatures. However, more recent research has focused on non-invasive methods of body temperature measurement for application in long-term monitoring. For example, thermal imaging captures the infrared radiation emitted by an individual, and is able to provide a detailed illustration of the body's heat distribution. Similarly, infrared thermometers detect infrared heat emitted at specific body sites such as the temporal artery [85].

Specific external measurement sites are known to provide accurate estimations of core body temperature. One such site is the tympanum, which receives a blood supply directly from the hypothalamus via the carotid artery [83]. However, representations of temperature distributions at the skin's surface are valuable as they reflect microcirculatory adjustments made in response to sympathetic stimuli.

For instance, digital temperature fluctuations are especially useful as indications of sympathetic nervous system activity because they have exclusive sympathetic innervation and dense cutaneous vasculature [86]. Transient skin temperature variations can be used to detect an individual's emotional state; for example, using infrared thermography, facial skin temperature is highly labile and has potential to serve as an access pathway in rehabilitation medicine [84].

3.4.3. Textile Innovations

Interested in determining optimal sensor placement for accurate mobile measurement of human body temperature, Li et al. [33] presented a novel wearable sensor and a finite element model of body temperature. Sensors were located over the left and right chest, at the axilla and upper back. The sensors consisted of short optical fibers embedded in polymer resin strips and based on Bragg grating (FBG) principles, reflecting light only at a specific wavelength while transmitting all others. The FBG wavelength changed linearly with temperature and the fiber optics were easily integrable in textiles. Although it is not clear whether or not a final device was tested on human subjects, the authors developed a weighted model of the five measurement locations to derive estimates of human body temperature, reportedly, to within 0.1 °C.

Shifting the focus away from the sensors themselves, Zysset et al. [48] introduced methods of better integrating electronics with textiles via gold- or copper-coated flexible plastic strips, which can act as a platform for surface mount devices (SMDs). Three methods of integration were investigated, namely, (a) wrapping plastic strips in cotton yarn using a roving frame; (b) weaving plastic strips into textiles; and (c) embroidering strips onto the fabric surface. Mechanical strain in the sensors was tested and embroidery was found to introduce the least strain into the plastic strips during fabrication, which is important to maintain the integrity of the circuits and any solder used to attach SMDs. A resistive temperature detector (RTD) consisting of a 100 nm thick gold layer on a plastic substrate exhibited a linear response between 30 °C and 90 °C as measured in a climate chamber. The response was not adversely affected by the weaving process.

3.4.4. Merits and Limitations

While significant advancements have been made in the integration of temperature sensors for non-invasive monitoring, these novel sensors have not yet been tested on human subjects. The sensors perform well theoretically and under controlled conditions, but without human testing it is impossible to determine whether they will be robust and accurate enough for long-term ambulatory monitoring. That being said, clinical methods of temperature measurement can be relatively non-invasive and there is limited evidence to suggest that long-term daily monitoring of temperature is valuable, making the intended application of a wearable temperature monitor unclear. Although Zysset et al. [48] address the issue of strain during manufacturing, it is unclear how fabric-based temperature sensors will respond to the stress of repeated washing and wearing.

3.5. Electrodermal Activity

3.5.1. Phenomenological Background

Electrodermal activity (EDA) is a measure of the skin's electrical conductivity. Eccrine sweat glands, found throughout most skin tissue, secrete a dilute electrolytic solution composed of primarily water and sodium chloride. The presence of these conducting solutes on the skin surface gives dermal conductivity, which varies with sweat gland activity. Electrodermal responses (EDRs) are defined as changes in EDA of over 0.05 μS within five seconds and are useful indicators of sweat gland activation [87].

Sweat glands are innervated by the sympathetic branch of the autonomic nervous system and fill with electrolytic fluid upon cholinergic stimulation. The subsequent secretion of these conducting solutes changes the skin's overall electrical conductance. Thus, the skin's resistive properties reflect the level and extent of an individual's sympathetic activity [31]. Homeostatic and thermoregulatory processes produce longer-term fluctuations in EDA, establishing a baseline upon which short-term sympathetically activated EDRs are superimposed.

3.5.2. Current Practice

Similar to bioimpedance, silver-silver chloride (Ag/AgCl) electrodes are commonly used to measure EDA. EDA measurement uses Ohm's Law (I = V/R) to calculate the electrical resistance at the skin surface. Figure 4 is a model of the electrode-skin interface for electrodermal measurements [88]. Vsd is a potential between the lumen of the sweat duct and the dermis and subcutaneous layers of skin. The parallel Rsd-Csd represents the resistance and capacitance of the wall of the sweat gland and duct. The remaining components of the model are the same as those in Figure 2.

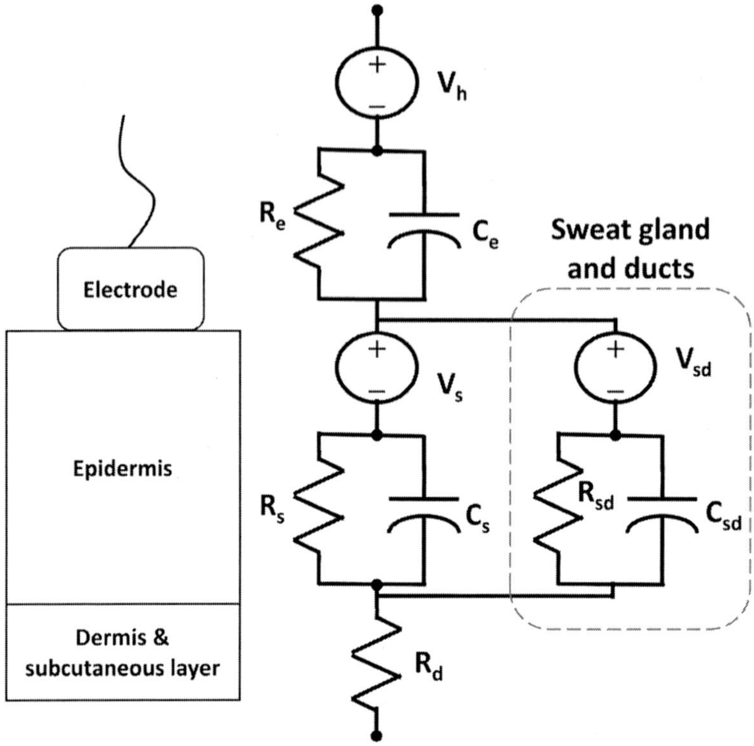

Figure 4. Electrical model of electrode-skin interface for electrodermal measurements.

The most well-established method of EDA measurement entails the exosomatic detection of current (I) when a constant voltage (V) is applied between two electrodes [89]. Less commonly, a constant current will be applied while voltage is measured to quantify the resistance (R) between the two electrodes. Ideally, electrodes are placed in regions of high sweat gland density where the greatest relative changes in EDA will take place. Typically the palms of the hands or the soles of the feet are the preferred locations [90]. EDRs can be interpreted as measures of physiological and psychological arousal, but they can be indicative of both negative and positive stimuli [89]. In clinical rehabilitation, electrodermal measurements may serve as a complementary access pathway to a motor-based switch [91] or brain-computer interface [92], an indicator of autonomic nervous system response to anxiety in children with autism spectrum disorders [93], and a quantitative measure of motor imagery abilities in individuals with spinal cord injuries [94], among other possibilities.

3.5.3. Textile Innovations

Despite the aforementioned applications in clinical rehabilitation, measurement of electrodermal activity is challenging outside of a controlled environment; typical sensors (wired electrodes on the fingertips) preclude bimanual activity, are susceptible to motion artefact and are sensitive to changes in ambient humidity and temperature. Textile sensing may circumvent some of these limitations and offer long-term, continuous electrodermal measurement.

With the goal of detecting drowsiness, Lee et al. [95] proposed a glove and armband system with textile-based EDA sensors in the fingers and pulse wave sensors in the wrist. Although the authors observed the expected increase in skin impedance and decrease in the phasic components of the electrodermal response in fatigue conditions, the ability of the system to automatically distinguish between states was not demonstrated quantitatively. Conversely, Lanata et al. [28] were able to distinguish between 5 arousal states with high accuracies (>88) using a glove system containing textile sensors at the index and middle fingertips. Sensors were a blended knit, composed of 80% polyester yarn and 20% stainless steel yarn, and were used simultaneously with standard Ag/AgCl electrodes while subjects observed pictures from the international affective picture system (IAPS), designed to elicit emotional arousal. Correlation between textile and standard electrodes was high (Spearman correlation > 0.95) for both tonic and phasic components of skin conductance.

Textile sensing researchers have also explored the ability to reliably measure EDA from locations other than the palm or fingers. Specifically,

Fletcher et al. [20] investigated the use of electrodes made of conductive knit fabric (silver-plated nylon blend) to measure EDA from sensors worn on the wrist. Along with traditional sensors to measure photoplethysmogram, temperature, and body motion, the purpose of the low cost, low power "iCalm" device was to identify and measure autonomic arousal in individuals with autism spectrum disorder (ASD), who may be unable to recognize or communicate symptoms of anxiety. The textile-based measurements exhibited a similar phasic response to that observed with the widely used FlexComp system (thoughttechnology.com), although the magnitude of the response was generally attenuated. Textile measurements also exhibited a large baseline drift, which the authors attributed to an insufficient pre-measurement time for skin-electrode interface stabilization. Consequently, they recommended a minimum wait of 15 min before measurement in future studies but did not present the longitudinal data on which this temporal threshold was based.

3.5.4. Merits and Limitations

To date, long-term unobtrusive measurement of EDA has been inconvenient at best, due to the primary sources of reliable response being the palm or fingertips. Recent research has focused on obtaining reliable measurement from other locations (such as the wrists as discussed above [20] and the toes and feet [90]) to improve wearability, but researchers have not yet been able to demonstrate the ability of textile-based devices to gather and analyze EDA signals with sufficient accuracy and repeatability to offer useful feedback to the wearer or remotely to a health professional. Although advances have been made in form factor, cost, and power consumption (e.g., the iCalm device appears smaller, more discreet, and more accurate than the device produced by Lee et al. [95]), quantitative data are needed to establish the clinical relevance of such devices.

3.6. Miscellaneous

Several other papers were found during the review process but fell outside of the above phenomenological categories. Some focused on system [16,47] or garment characteristics [25,40], while one paper examined applications in actuation as opposed to simple monitoring [27] and one explored textile sensing of electroencephalography (EEG) [34].

Angelidis [16] presented a broad overview of a system including electronic textiles and biosensors for vital physical monitoring during physical activity. The conceptual paper gave an overview of an idealized system and identified steps and current barriers to its realization, including signal quality, miniaturization and packaging. Zheng et al. [47] argued that battery life is a determining factor in the usability of wearable technologies and presented a novel battery-scheduling algorithm to extend the operation of multi-battery electronic textile garments.

Schwarz et al. [40] tackled common criticisms of electronic fabrics and yarns by developing yarns with improved elasticity, drape, and mechanical strength. While the authors developed an electrical model to predict the properties of various yarns, an analysis of elasticity and drape was not provided and no recommendations were made about preferred material characteristics.

Specifically targeted at electrical stimulation to alleviate hypertension in older adults, Kim and Cho [27] presented the only paper focused on both monitoring and treatment. A glove-based system was used to sense blood pressure and heart rate in twelve elderly female patients before and after electrical stimulation was delivered through stainless steel thread within the glove. Rings of conductive thread, 20 mm in diameter, were stitched and used to deliver user-controlled electrical stimulation between 1 to 1200 Hz for 15 min. Decreased blood pressure ($p < 0.01$) was reported following 15 min of glove wear.

Lofhede et al [34] suggest the use of textile electrodes to facilitate long-term recording of EEG in pre-term infants without discomfort or damage to the scalp. The authors investigated two types of textile electrodes, one made from a silver plated elastomer blend, the other knit with nylon, spandex, and conductive silver fibers. When a conductive gel or saline was used as a contact medium, signals from both types of textile electrode were similar to those recorded from standard electrodes at the same locations on the scalp. Although signal quality was deemed to be acceptable in both cases, the five study participants reported no benefits in terms of comfort of textile over standard electrodes. Textile electrodes may prove to be more comfortable in longer term testing, but additional research is necessary to verify the signal quality and diagnostic validity of textile electrodes for EEG.

4. DISCUSSION

In e-textiles research, the ultimate objective is usually full integration, with sensing elements and electronics indistinguishable from everyday clothing. While research and technology has not yet advanced to this point, significant progress has been made toward this goal since the early prototypes of the Wearable Motherboard in 2002 [1]. Primary advances in e-textiles have been in healthcare and monitoring applications. Large multi-lab studies like ProeTEX [96] and MyHeart [97] have led the field in integration. Signal quality has generally been the primary consideration while user comfort and sensor integration have been secondary. Despite significant advances in recent years, circuit and battery integration are still areas in need of additional improvement. As stated in an early review of e-textile technologies by Post et al. [63], integration can only be achieved by departing from the notion of "packaging electronics in hard plastic boxes, however small" [4]. Current research seems to be addressing these limitations with commercially available flexible circuit boards (www.ictboards.com) and batteries (www.solicore.com), although these technologies have been slow to make their way into smart garments, presumably because of the relatively high cost of these flexible components, at the time of writing.

There seem to be other significant gaps between technology advancement and implementation. In general, in e-textile research, there are many studies describing monitoring applications, but few on actuation [14]. Actuation and user feedback through visual effects, sound and even morphological changes are popular in fashion and DIY applications [2]. Although there is much interest in personal information tracking and biofeedback in our current culture, as evidenced by the growing array of wearable technology, this interest has yet to diffuse into the clinical realm. While there is a growing demand for more personal control of health information [16], there are an increasing number of studies where information is sent directly to a coach, clinician, healthcare professional, or simply into a database of other biosignals, without any immediate and direct feedback to the user of the technology. This approach seems to contradict the concepts of patient self-management [98].

In a 2001 review of wearable technology, concerns were raised about "little published data from field trials" [99], a concern that persists to the present day. While research with healthy subjects produces valuable initial data, research with clinical populations is a necessary next step as a number of real-life challenges cannot be simulated or even anticipated in the lab. Ethical concerns of fully integrated textile measurements as well as issues of participant compliance must be addressed before these technologies can move from research into practice. Privacy concerns have also been raised around the storage and transfer of moment-by-moment personal health information, with increasing security measures being implemented [13].

To accelerate future developments, the standardization of interconnections between e-textile elements may be necessary. Upon future mass production, e-textile waste products may also require novel recycling and reuse strategies. While we seek ubiquitous computing, many people remain wary of increasingly invisible technology allowing them to be unknowingly monitored or recorded. However, with mainstream products like Google Glass and Microsoft's HoloLens, these attitudes may be on the decline.

E-textiles are a comparatively young field with much room for further research and application. However, before these technologies are integrated into clinical rehabilitation practice, there are outstanding technological and ethical hurdles that must be addressed.

5. RECOMMENDATIONS

In light of the above review, we close with the following recommendations for future research.

1. Future research ought to validate textile sensing of a particular physiological or biomechanical phenomenon against its corresponding clinical gold standard. This was a common gap across the reviewed papers.
2. In the spirit of patient self-management, future work may entertain the potential of incorporating textile actuation and hence, immediate sensory feedback to the wearer.
3. While sensing elements have experienced a boon in fabric integration, connecting circuitry and power sources still lag behind in textile assimilation. Truly imperceptible e-textiles will require full system integration.
4. With continued improvements in e-textile signal quality and system integration, it would behoove researchers to initiate testing with clinical populations in ambulatory settings. In particular, future research should consider issues of signal stability over time and across user activities as well as textile sensor integrity with wear and wash.

6. LIMITATIONS

Although interesting research has no doubt been documented in conference papers and private or commercial documents, due to the sheer amount of articles published in recent years, we limited the scope of this review to peer-reviewed journal articles. As a consequence, some relevant advancements may have been omitted.

CONCLUSIONS

E-textiles constitute an exciting and growing field. Advancements in recent years have brought us closer to the vision of ubiquitous computing, particularly in the fields of remote and ambulatory monitoring in clinical rehabilitation. Many physiological and biomechanical phenomenons can already be measured via textile sensors. Nonetheless, there remain many opportunities for future research such as the integration of power and information processing components, systematic validation against clinical gold standards and sensory feedback via textile actuation.

CONFLICTS OF INTEREST

The authors declare no conflict of interest.

AUTHOR CONTRIBUTIONS

A.F. conceived of and conducted the review and critical literature appraisal; T.C. contributed to the writing and editing of various sections; M.S. assisted with the writing of the phenomenological backgrounds of various sections.

REFERENCES

1. Park, S.; Mackenzie, K.; Jayaraman, S. The Wearable Motherboard: A Framework for Personalized Mobile Information Processing (PMIP). In Proceedings of the 39th Design Automation Conference, New Orleans, LA, USA, 10–14 June 2002.

2. Buechley, L.; Eisenberg, M. Fabric PCBs, electronic sequins, and socket buttons: Techniques for e-textile craft. Pers. Ubiquitous Comput. 2009, 13, 133–150.

3. Meoli, D.; May-Plumlee, T. Interactive electronic textile development: A review of technologies. J. Textile Appar. Technol. Manag. 2002, 2, 1–12.

4. Post, E.R.; Orth, M.; Russo, R.R.; Gershenfeld, N. E-broidery: Design and fabrication of textile-based computing. IBM Syst. J. 2000, 39, 840–860.

5. Cottet, D.; Grzyb, J.; Kirstein, T.; Tröster, G. Electrical characterization of textile transmission lines. IEEE Trans. Adv. Packag. 2003, 26, 182–190.

6. Maggie, O. Defining Flexibility and Sewability in Conductive Yarns. In Proceedings of the Materials Research Society (MRS) Symposium, San Francisco, CA, USA, 25–29 April 2011; Volume 736, pp. 37–48.

7. Bonderover, E.; Wagner, S. A woven inverter circuit for e-textile applications. IEEE Electr. Device Lett. 2004, 25, 295–297.

8. Lee, J.B.; Subramanian, V. Organic Transistors on Fiber: A First Step towards Electronic Textiles. In Proceedings of the Technical Digest—International Electron Devices Meeting, Washington, WA, USA, 8–10 December 2003.

9. Lee, J.B.; Subramanian, V. Weave patterned organic transistors on fiber for e-textiles. IEEE Trans. Electr. Devices 2005, 52, 269–275.

10. Edmison, J.; Jones, M.; Nakad, Z.; Martin, T. Using piezoelectric materials for wearable electronic textiles. In Proceedings of the Sixth International Symposium on Wearable Computers (ISWC), Seattle, WD, USA, 7–10 October 2002; pp. 41–48.

11. Pacelli, M.; Loriga, G.; Taccini, N.; Paradiso, R.M.D. 3rd IEEE EMBS International Summer School on, B. Sensing fabrics for monitoring physiological and biomechanical variables: E-textile solutions. In Proceedings of the 3rd IEEE/EMBS International Summer School on Medical Devices and Biosensors, Cambridge, MA, USA, 4–6 September 2006.

12. Pantelopoulos, A.; Bourbakis, N.G. A survey on wearable sensor-based systems for health monitoring and prognosis. IEEE Trans. Syst. Man Cybern. Part C Appl. Rev. 2010, 40, 1–12.

13. Chan, M.; Esteve, D.; Fourniols, J.Y.; Escriba, C.; Campo, E. Smart wearable systems: Current status and future challenges. Artif. Intell. Med. 2012, 56, 137–156.

14. Bonato, P. Advances in wearable technology and applications in physical medicine and rehabilitation. J. NeuroEng. Rehabil. 2005, 2.

15. Adnane, M.; Jiang, Z.; Choi, S.; Jang, H. Detecting specific health-related events using an integrated sensor system for vital sign monitoring. Sensors 2009, 9, 6897–6912.

16. Angelidis, P.A. Personalised physical exercise regime for chronic patients through a wearable ICT platform. Int. J. Electron. Healthc. 2010, 5, 355–370.

17. Baek, H.J.; Chung, G.S.; Kim, K.K.; Park, K.S. A smart health monitoring chair for nonintrusive measurement of biological signals. IEEE Trans. Inf. Technol. Biomed. 2012, 16, 150–158.

18. Bianchi, A.M.; Mendez, M.O.; Cerutti, S. Processing of signals recorded through smart devices: Sleep-quality assessment. IEEE Trans. Inf. Technol. Biomed. 2010, 14, 741–747.

19. Cho, H.S.; Koo, S.M.; Lee, J.; Cho, H.; Kang, D.H.; Song, H.Y.; Lee, J.W.; Lee, K.H.; Lee, Y.J. Heart monitoring garments using textile electrodes for healthcare applications. J. Med. Syst. 2011, 35, 189–201.

20. Fletcher, R.R.; Dobson, K.; Goodwin, M.S.; Eydgahi, H.; Wilder-Smith, O.; Fernholz, D.; Kuboyama, Y.; Hedman, E.B.; Poh, M.Z.A.; Picard, R.W.A. ICalm: Wearable sensor and network architecture for wirelessly communicating and logging autonomic activity. IEEE Trans Inf. Technol Biomed. 2010, 14, 215–223.

21. Gioberto, G.; Dunne, L.E. Overlock-stitched stretch sensors: Characterization and effect of fabric property. J. Textile Appar. Technol. Manag. 2013, 8, 1–14.

22. Giorgino, T.; Tormene, P.; Lorussi, F.; de Rossi, D.; Quaglini, S. Sensor evaluation for wearable strain gauges in neurological rehabilitation. IEEE Trans. Neural Syst. Rehabil. Eng. 2009, 17, 409–415.

23. Goy, C.B.; Dominguez, J.M.; Gómez López, M.A.; Madrid, R.E.; Herrera, M.C. Electrical characterization of conductive textile materials and its evaluation as electrodes for venous occlusion plethysmography. J. Med. Eng. Technol. 2013, 37, 359–367.

24. Hannikainen, J.; Vuorela, T.; Vanhala, J. Physiological measurements in smart clothing: A case study of total body water estimation with bioimpedance. Trans. Inst. Meas. Control 2007, 29, 337–354.

25. Harms, H.; Amft, O.; Troester, G. Estimating posture-recognition performance in sensing garments using geometric wrinkle modeling. IEEE Trans. Inf. Technol. Biomed. 2010, 14, 1436–1445.

26. Hong, S.; Yang, Y.; Kim, S.; Shin, S.; Lee, I.; Jang, Y.; Kim, K.; Yi, H. Performance study of the wearable one-lead wireless electrocardiographic monitoring system. Telemed. J. E-Health 2009, 15, 166–175.

27. Kim, R.H.; Cho, G. Effectiveness of the smart healthcare glove system for elderly persons with hypertension. Hum. Factors Ergon. Manuf. 2013, 23, 198–212.

28. Lanata, A.; Gaetano, V.; Scilingo, E. A novel EDA glove based on textile-integrated electronics for affective computing. Med. Biol. Eng. Comput. 2012, 50, 1163–1172.

29. Lee, S.; Ahn, Y.; Prabu, A.; Kim, K. Piezoelectric Polymer and Piezocapacitive Nanoweb Based Sensors for Monitoring Vital Signals and Energy Expenditure in Smart Textiles. J. Fiber Bioeng. Inf. 2013, 6, 369–381.

30. Lee, B.W.; Lee, C.; Kim, J.; Lee, M. Optimum conductive fabric sensor sites for evaluating the status of knee joint

movements using bio-impedance. Biomed. Eng. Online 2011, 10, 48.

31. Fowles, D.; Coles, M.; Donchin, E.; Porges, S. The eccrine system and electrodermal activity. In Psychophysiology; Guildford Press: New York, NY, USA, 1985; pp. 51–96.

32. Lee, Y.D.; Chung, W.Y. Wireless sensor network based wearable smart shirt for ubiquitous health and activity monitoring. Sens. Actuators B Chem. 2009, 140, 390–395.

33. Li, H.; Yang, H.; Li, E.; Liu, Z.; Wei, K. Wearable sensors in intelligent clothing for measuring human body temperature based on optical fiber Bragg grating. Opt. Express 2012, 20, 11740–11752.

34. Lofhede, J.; Seoane, F.; Thordstein, M. Textile electrodes for EEG recording—A pilot study. Sensors 2012, 12, 16907–16919.

35. Lopez, G.; Custodio, V.; Ignacio Moreno, J. LOBIN: E-textile and wireless-sensor-network-based platform for healthcare monitoring in future hospital environments. IEEE Trans. Inf. Technol. Biomed. 2010, 14, 1446–1458.

36. Lorussi, F.; Galatolo, S.; Bartalesi, R. Modeling and characterization of extensible wearable textile-based electrogoniometers. IEEE Sens. J. 2013, 13, 217–228.

37. Marquez, J.C.; Seoane, F.; Lindecrantz, K. Textrode functional straps for bioimpedance measurements - experimental results for body composition analysis. Eur. J. Clin. Nutr. 2013, 67 (Suppl. 1), S22–S27.

38. Preece, S.J.; Kenney, L.P.; Major, M.J.; M.J., T.; Lay, E.; Fernandes, B.T. Automatic identification of gait events using an instrumented sock. J. NeuroEng. Rehabil. 2011, 8, 32.

39. Di Rienzo, M.; Rizzo, F.; Meriggi, P.; Castiglioni, P.; Mazzoleni, P.; Parati, G.; Bordoni, B.; Brambilla, G.; Ferratini, M. MagIC system. IEEE Eng. Med. Biol. Mag. 2009, 28, 35–40.

40. Schwarz, A.; Cuny, L.; Hertleer, C.; Ghekiere, F.; Kazani, I.; de Clercq, G.; de Mey, G.; van Langenhove, L.A. Electrical circuit model of elastic and conductive yarns produced by hollow spindle spinning. Mater. Technol. 2011, 26, 121–127.

41. Shu, L.; Hua, T.; Wang, Y.; Li, Q.; Feng, D.D.; Tao, X. In-shoe plantar pressure measurement and analysis system based on fabric pressure sensing array. IEEE Trans. Inf. Technol. Biomed. 2010, 14, 767–775.

42. Song, H.Y.; Lee, J.H.; Kang, D.; Cho, H.; Cho, H.S.; Lee, J.W.; Lee, Y.J. Textile electrodes of jacquard woven fabrics for biosignal measurement. J. Textile Inst. 2010, 101, 758–770.

43. Tormene, P.; Bartolo, M.; De Nunzio, A.M.; Fecchio, F.; Quaglini, S.; Tassorelli, C.; Sandrini, G. Estimation of human trunk movements by wearable strain sensors and improvement of sensor's placement on intelligent biomedical clothes. Biomed. Eng. Online 2012, 11, 95.

44. Vuorela, T.; Seppa, V.P.; Vanhala, J.; Hyttinen, J. Design and implementation of a portable long-term physiological signal recorder. IEEE Trans. Inf. Technol. Biomed. 2010, 14, 718–725.

45. Yamada, T.; Hayamizu, Y.; Yamamoto, Y.; Yomogida, Y.; Izadi-Najafabadi, A.; Futaba, D.N.; Hata, K. A stretchable carbon nanotube strain sensor for human-motion detection. Nat. Nanotechnol. 2011, 6, 296–301.

46. Zhang, Z.B.; Shen, Y.H.; Wang, W.D.; Wang, B.Q.; Zheng, J.W. Design and implementation of sensing shirt for ambulatory cardiopulmonary monitoring. J. Med. Biol. Eng. 2011, 31, 207–216.

47. Zheng, N.; Wu, Z.; Lin, M.; Yang, L.T. Enhancing battery efficiency for pervasive health-monitoring systems based on electronic textiles. IEEE Trans. Inf. Technol. Biomed. 2010, 14, 350–359.

48. Zysset, C.; Kinkeldei, T.; Münzenrieder, N.; Petti, L.; Salvatore, G.; Tröster, G. Combining electronics on flexible plastic strips with textiles. Textile Res. J. 2013, 83, 1130–1142.

49. Mirvis, D.; Goldberger, A. Electrocardiography. In Braunwald's Heart Disease, 10th ed.; Saunders Elsevier: Philadelphia, PA, USA, 2015; pp. 114–154.

50. Camm, A.; Malik, M.; Bigger, J.; Breithardt, G.; Cerutti, S.; Cohen, R.; Coumel, P.; Fallen, E.; Kennedy, H.; Kleiger, R.; et al. Heart rate variability: standards of measurement, physiological interpretation and clinical use. Task Force of the European Society of Cardiology and the North American Society of Pacing and Electrophysiology. Circulation 1996, 93, 1043–1065.

51. Neumann, M. Biopotential electrodes. In Clinical Engineering; David, Y., von Maltzahn, W., Neuman, M., Bronzino, J., Eds.; CRC Press: Boca Raton, FL, USA, 2003; pp. 10:1–10:12.

52. Webster, J. Medical Instrumentation: Application and Design, 4th ed.; Wiley: Hoboken, NJ, USA, 2010.

53. Su, L.; Borov, S.; Zrenner, B. 12-lead Holter electrocardiography: Review of the literature and clinical application update. Herzschrittmachertherapie Elektrophysiol. 2013, 24, 92–96.

54. Acharya, U.R.; Joseph, K.P.; Kannathal, N.; Lim, C.M.; Suri, J.S. Heart rate variability: A review. Med. Biol. Eng. Comput. 2006, 44, 1031–1051.

55. Appelhans, B.; Luecken, L. Heart rate variability as an index of regulated emotional responding. Rev. General Psychol. 2006, 10, 229–240.

56. Coyle, M. Ambulatory cardiopulmonary data capture. In Proceedings of the 2nd Annual International IEEE-EMBS Special Topic Conference on Microtechnologies in Medicine & Biology, Madison, WI, USA, 2–4 May 2002; pp. 297–300.

57. Coosemans, J.; Hermans, B.; Puers, R. Integrating wireless ECG monitoring in textiles. Sens. Actuators A 2006, 130–131, 48–53.

58. Lim, Y.; Kim, K.; Park, K. ECG recording on a bed during sleep without direct skin-contact. IEEE Trans. Biomed. Eng. 2007, 54, 718–725.

59. Grimnes, S.; Martinsen, O. Bioimpedance Biolectricity Basics, 2nd ed.; Academic: London, UK, 2008.

60. Kyle, U.; Bosaeus, I.; Lorenzo, A.D.; Deurenberg, P.; Elia, M.; Gomez, J.; Heitmann, B.; Kent-Smith, L.; Melchoir, J.; Pirlich, M.; et al. Bioelectrical impedance analysis—Part I: Review of principles and methods. Clin. Nutr. 2004, 23, 1226–1243.

61. Kushner, R. Bioelectrical impedance analysis: A review of principles and applications. J. Am. Coll. Nutr. 1992, 11, 199–209.

62. Togawa, T.; Tamura, T.; Oberg, P.A. Biomedical Transducers and Instruments; CRC Press: Boca Raton, FL, USA, 1997.

63. Cornish, B.; Jacobs, A.; Thomas, B.; Ward, L. Optimizing electrode sites for segmental bioimpedance measurements. Physiol. Meas. 1999, 20, 241–250.

64. Lescolarde, L.; Yanguas, J.; Lukaski, H.; Alomar, X.; Rosell-Ferrer, J.; Rodas, G. Localized bioimpedance to assess muscle injury. Physiol. Meas. 2013, 34, 237–245.

65. Kushner, R.F.; Schoeller, D.A. Estimation of total body water by bioelectrical impedance analysis. Am. J. Clin. Nutr. 1986, 44, 417–424.

66. Agarwal, R.; Yadav, R.; Anand, S.; Suri, J.C.; Girija, J. Electrical impedance plethysmography technique in estimating pulmonary function status. J. Med. Eng. Technol. 2007, 31, 1–9.

67. Azar, R.; al-Moubarak, I.; Barsumau, J.; Smessaert, C.; Vairon, M.X. Assessment and follow-up of nutritional status in hemodialysis patients. Néphrol. Thér. 2009, 5, S317–S322.

68. Medrano, G.; Eitner, F.; Floege, J.; Leonhardt, S. A novel bioimpedance technique to monitor fluid volume state during hemodialysis treatment. ASAIO J. 2010, 56, 215–220.

69. Sjedic, E.; Fu, Y.; Pak, A.; Fairley, J.; Chau, T. The effects of rhythmic sensory cues on the temporal dynamics of human gait. PLoS One 2012, 7, e43104.

70. Masani, K.; Vette, A.; Popovic, M. Controlling balance during quiet standing: proportional and derivative controller generates preceding motor command to body sway position observed in experiments. Gait Posture 2006, 23, 164–172.

71. Lesperance, A.; Blain, S.; Chau, T. An integrated approach to detecting communicative intent amid hyperkinetic movements in children. Augment. Altern. Commun. 2011, 27, 150–162.

72. Tong, L.; Song, Q.; Ge, Y.; Liu, M. HMM-based human fall detection and prediction method using tri-axial accelerometer. IEEE Sens. J. 2013, 13, 1848–1856.

73. Gibbs, P.T.; Asada, H.H. Wearable conductive fiber sensors for multi-axis human joint angle measurements. J. NeuroEng. Rehabil. 2005, 2, 18.

74. Edmison, J.; Jones, M.; Lockhart, T.; Martin, T. An e-textile system for motion analysis. Stud. Health Technol. Inform. 2004, 108, 292–301.

75. Dunne, L.E.; Walsh, P.; Hermann, S.; Smyth, B.; Caulfield, B. Wearable monitoring of seated spinal posture. IEEE Trans. Biomed. Circuits Syst. 2008, 2, 97–105.

76. Kuang, K.; Cantwell, W.J. An evaluation of a novel plastic optical fibre sensor for axial strain and bend measurements. Meas. Sci. Technol. 2002, 13, 1523.

77. Lorussi, F.; Galatolo, S.; de Rossi, D.E. Textile-based electrogoniometers for wearable posture and gesture capture systems. IEEE Sens. J. 2009, 9, 1014–1024.

78. Hall, M.A.; Holmes, G. Benchmarking attribute selection techniques for discrete class data mining. IEEE Trans. Knowl. Data Eng. 2003, 15, 1437–1447.

79. Gioberto, G.; Dunne, L. Theory and characterization of a top-thread coverstitched stretch sensor. In Proceedings of the IEEE International Conference on Systems, Man and Cybernetics, Seoul, Korea, 14–17 October 2012; pp. 3275–3280.

80. Sund-Levander, M.; Forsberg, C.; Wahren, L. Normal oral, rectal, tympanic and axillary body temperature in adult men and women: A systematic literature review. Scand. J. Caring Sci. 2002, 16, 122–128.

81. Charkoudian, N. Skin blood flow in adult human thermoregulation: How it works, when it does not and why. Mayo Clin. Proc. 2003, 78, 603–612.

82. Oberle, J.; Elam, M.; Karlsson, T.; Wallin, B. Temperature-dependent interaction between vasoconstrictor and vasodilator mechanisms in human skin. Acta Physiol. Scand. 1988, 132, 459–469.

83. Sund-Levander, M.; Grodzinsky, E. Assessment of body temperature measurement options. Br. J. Nurs. 2013, 22, 944–950.

84. Nhan, B.; Chau, T. Classifying affective states using thermal infrared imaging of the human face. IEEE Trans. Biomed. Eng. 2010, 57, 979–987.

85. Allegaert, K.; Casteels, K.; van Gorp, I.; Bogaert, G. Tympanic, infrared skin and temporal artery scan thermometers compared with rectal measurement in children: A real-life assessment. Curr. Ther. Res. 2014, 76, 34–38.

86. Keefe, F.; Gardner, E. Learned control of skin temperature: Effects of short- and long-term biofeedback training. Behav. Ther. 1979, 10, 202–210.

87. Blain, S.; Mihailidis, A.; Chau, T. Assessing the potential of electrodermal activity as an alternative access pathway. Med. Eng. Phys. 2008, 30, 498–505.

88. Boccaletti, C.; Castrica, F.; Fabbri, G.; Santello, M. A non-invasive biopotential electrode for the correct detection of bioelectric currents. In Proceedings of the Sixth IASTED International Conference on Biomedical Engineering, Innsbruck, Austria, 13–15 February 2008; pp. 353–358.

89. Blain, S.; Chau, T.; Mihailidis, A. Peripheral autonomic signals as access pathways for individuals with severe disabilities: A literature appraisal. Open Rehabil. J. 2008, 1, 27–37.

90. Kushki, A.; Fairley, J.; Merja, S.; King, G.; Chau, T. Comparison of blood volume pulse and skin conductance responses to mental and affective stimuli at different anatomical sites. Physiol. Meas. 2011, 32, 1529–1539.

91. Leung, B.; Chau, T. Autonomic responses to correct outcomes and interaction errors during single-switch scanning among children with severe spastic quadriplegic cerebral palsy. J. NeuroEng. Rehabil. 2014, 11, 13.

92. Falk, T.; Guirgis, M.; Power, S.; Chau, T. Taking NIRS-BCI outside the lab: towards achieving robustness against environmental noise. IEEE Trans. Neural Syst. Rehabil. Eng. 2010, 19, 136–146.

93. Kushki, A.; Drumm, E.; Mobarak, M.P.; Tanel, N.; Dupuis, A.; Chau, T.; Anagnostou, E. Investigating the autonomic

nervous system response to anxiety in children with autism spectrum disorders. PLoS One 2013, 8, e59730.

94. Grangeon, M.; Charvier, K.; Guillot, A.; Rode, G.; Collet, C. Using sympathetic skin responses in individuals with spinal cord injury as a quantitative evaluation of motor imagery abilities. Phys. Ther. 2012, 92, 831–840.

95. Lee, Y.; Lee, B.; Lee, M. Wearable sensor glove based on conducting fabric using electrodermal activity and pulse-wave sensors for e-health application. Telemed. J. E-Health 2010, 16, 209–217.

96. Curone, D.; Secco, E.L.; Tognetti, A.; Loriga, G.; Dudnik, G.; Risatti, M.; Whyte, R.; Bonfiglio, A.; Magenes, G. Smart garments for emergency operators: The ProeTEX project. IEEE Trans. Inf. Technol. Biomed. 2010, 14, 694–701.

97. Habetha, J. The MyHeart project-fighting cardiovascular diseases by prevention and early diagnosis. In Proceedings of the IEEE EMBS Annual International Conference, New York, NY, USA, 30 August–3 September 2006; pp. 6746–6749.

98. Shumaker, S.; Ockene, J.; Riekert, K. The Handbook of Health Behavior Change, 3rd ed.; Springer Publishing Company: New York, NY, USA, 2009.

99. Baber, C. Wearable computers: A human factors review. Int. J. Hum. Comput. Interact. 2001, 13, 123–145.

CHAPTER 2

Electro-Conductive Sensors and Heating Elements Based on Conductive Polymer Composites in Woven Fabric Structures

Irina Cristian[1, 2, 3], Saad Nauman[1, 2], Cédric Cochrane[1, 2], Vladan Koncar[1, 2]

[1] *University Lille Nord de France, France*

[2] *ENSAIT, GEMTEX, F-59100 Roubaix, France*

[3] *Technical University GHEORGHE ASACHI of Iaşi , Romania*

1. INTRODUCTION

The use of intelligent materials which are capable of reacting to external stimuli is growing in the field of textiles. Conductive materials, metals as well as conducting polymers, are already being used in many textile applications, like antistatic materials, electromagnetic interference shielding, heating, transport of electrical signals, sensors etc.

Conductive polymers are some of the developments which seem to respond to specific properties of textile materials, like flexibility and deformability. Two sub-classes of conductive polymers can be identified: intrinsically conductive polymers and conductive polymer composites.

Inherently conducting polymers are composed of polymer chains containing long conjugated double bonds which give rise to highly conducting properties, and they have been considered as promising materials (Heeger, 2002; Kumar et al., 1998). Inherently conducting polymers are suitable for applications in many domains of intelligent textiles, but they present some substantial disadvantages, like infusibility or insolubility in common organic solvents, weak mechanical properties and poor procesability. However, advanced solution processing of this class of polymers has been developed significantly over the last decade to improve their low solubility. For example, chemical modification of monomers with dopants has enhanced the solubility in the case of polythiophene and polyaniline (Haba et al., 1999; Gettinger et al., 1995).

For their part, composite conductive polymers are obtained by blending (generally by melt mixing) an insulating polymer matrix (thermoplastic or thermosetting plastic) with conductive fillers like carbon black, carbon fibres or nanotubes, metallic particles or conductive polymers. The presence of filler particles in the matrix may have a negative impact on the mechanical properties of the final composite (Krupa et al., 2001; Novak et al., 2002). Instead of this, the development in the field of composite conductive polymers seems therefore to be a promising approach for intelligent textiles own to simplicity of preparing and to their low cost.

In this chapter, two applications (sensor and actuator) based on coating of textile materials with conductive polymer composites are presented. Several different coating techniques for intelligent textile structures exist - the one chosen for both applications was developed in our laboratory (Cochrane et al., 2007, 2010) based on dispersed carbon black particles (Printex® L6) in a polymer solution (Styren-Butadien-Styren or latex), using a solvent.

In the first part of this chapter, a new approach of NDE (Non Destructive Evaluation) using fibrous sensors inserted inside composite woven reinforcements during their weaving is presented. The use of 3D woven fabrics as the reinforcing medium for composites is becoming a popular choice, due to various advantages such as reduced cost and shorter production cycle, greater design flexibility and superior mechanical properties (Kamiya et al., 2000). Recently, these high performance composites reinforced with 3D structures have found wide applications in various industrial areas such as aerospace, aircraft, automobile, civil engineering etc. Good quality and reliability are basic requirements for advanced composite structures which are often used under harsh environments. To improve their performance, the cure monitoring of technological process is clearly necessary. At the same time, in service NDE is also needed to keep these structures operating safely and reliably.

A novel flexible piezoresistive fibrous sensor has been developed and optimized for in situ structural deformation sensing in carbon composites. Those sensors were inserted as weft in 3D-woven interlock reinforcement, during the weaving process on a special weaving loom. The reinforcement was then impregnated in epoxy resin and was later subjected to quasi static tensile loading. It was found that the sensor was able to detect deformations in the composite structure until rupture since it was inserted together with reinforcing tows.

The morphological and electromechanical properties of the fibrous sensors have been analyzed using tomography and yarn tensile strength tester. An appropriate data acquisition module has also been developed and used for data acquisition and its further treatment. The results obtained for carbon composite specimens under standard testing conditions have validated in situ monitoring concept using our fibrous textile sensors.

The second part of this chapter presents a woven fabric containing an original heating element. Textile actuators like heating fabrics can find applications in numerous and varied fields such as sports, leisure, medical and automotive (Droval et al., 2005; El-Trantawy et al., 2002). Usually actuators need heavy power supplies that are rarely flexible and lightweight which badly affects wearability. Our heating element is designed to adapt to flexible structures. Metallic yarns, used as electrodes are integrated in a woven structure in a comb-teeth arrangement. These electrodes are connected to a power supply. A thin conductive coating is applied on the fabric surface and on the electrode arrangement, in order to ensure uniform heat distribution. The coating is a composite material based on aqueous latex dispersed with carbon black as filler. The heating element (comb electrodes and electro-conductive coating) can thus adopt the desired pattern. This is an important aspect of our heating element as it allows integration of the heating element in various fabrics designed for varied and diverse applications. In our research the distance between electrodes remained unchanged, the only variable value being the percentage of filler. The results regarding the power consumption and textile surface temperature for different filler content were registered. From these results, optimum filler content for heating application was defined. A thermal image of the heating fabric operating at low voltage is given, in order to demonstrate the efficiency of our system. A more precise measurement of surface temperature is thus possible. For applied voltage of 15 V, the maximum temperature gained was about 50 °C. The thermal image also demonstrates the homogeneity of heating provided by our system. Potential applications of these self heating fabrics include garments designed to provide thermal comfort and antifreeze safety.

2. ELECTRO-CONDUCTIVE SENSORS FOR ON-LINE MEASUREMENTS OF STRUCTURAL DEFORMATION IN COMPOSITES REINFORCED WITH 3D-WOVEN FABRICS

Weaving technique has been used for a long time in order to obtain technical textile products for industrial applications. An important use of this technology is for the manufacturing of 3D reinforcements using high performance fibres (carbon, glass, aramid etc.). 3D reinforcement based composites, in combination with high-performance fibres, are being increasingly used in the aerospace industry (Ko, 2007). Particular advantages of these fabrics mentioned in the literature include better through the thickness properties, better out of plane properties, high impact resistance, enhanced delamination resistance, resistance to crack propagation, impact/fracture resistance, improved post impact mechanical properties, damage tolerance, dimensional stability, ease of fabrication and minimal need of cutting, lay up and joining.

Regarding these properties it can be safely concluded that 3D woven fabrics constitute the most promising class of reinforcements for composite materials for high tech structural applications. To improve their performance, the cure monitoring of technological process is clearly necessary. At the same time, in-service non destructive evaluation is also needed to keep these structures operating safely and reliably. Non destructive evaluation techniques have been developed in the past including ultrasonic scanning, acoustic emission, shearography, stimulated infrared thermography, fibre brag grating sensors and vibration testing etc. (Black, 2009). The challenge today is to develop new low cost techniques which can perform on-line structural health assessment starting from the manufacture of composite structure to the real service of these structures in the field. Moreover, the non destructive evaluation techniques have to be integrated in the design phase and sensors should be inserted during the fabrication of composites, in order to improve accuracy and reduce their costs. The classical non destructive evaluation techniques are difficult to adapt. They are not well suited for on-line structural health monitoring, because of difficulties in making in situ implementation.

One possible solution is to use intelligent textile materials and structures which provide real possibility for on-line and in situ monitoring of structural integrity. Such intelligent materials are made by coating or treating textile yarns, filaments or fabrics with nanoparticles or conductive & semi-conductive polymers giving them special properties.

A review of piezoresistive sensing approaches already being applied to measure strain in fabrics/composites shows that several sensing mechanisms exist (Dharap et al., 2004; Lorussi et al., 2005; Scilingo et al., 2003; Fiedler et al., 2004). These approaches may be categorized on the basis of manufacturing technology as nanotube networks, use of carbon tows for self-sensing and semi-conductive coatings.

Nanotubes have been investigated in detail for use as sensing mechanisms, both for smart textile applications and for structural health monitoring of composites. Significant challenges still exist in their development, for example the efficient growth of macroscopic-length carbon nanotubes, controlled growth of nanotubes on desired substrates, durability of nanotube based sensors and actuators, effective dispersion in polymer matrices and their orientation. Therefore, there is a need to develop both experimental and analytical techniques to bridge the nano and macro scales towards optimization so as to use nanotube networks as sensors inside macroscale (fabric) or mesoscale (tow) composites (Li et al., 2005).

Carbon fibre reinforced composites offer a unique possibility of using carbon tows as sensing network because of their conductivity. However such an approach can only be used for conductive fibre based composites. Moreover, before applying such an approach for structural health monitoring it is imperative to understand the deformation mechanism of the reinforcement. Any anomaly in the deformation mechanism can threaten the sensing mechanism's validity and efficacy.

Concerning semi conductive coatings, they are easy to realize and can be made wash resistant. Their use as percolation networks for sensing in structural health monitoring applications is quite promising and needs to be further investigated.

Present study is aimed at designing, developing and optimizing piezoresistive fibrous sensors realized from semi-conductive coatings, suited for composites structural parts containing 3D reinforcement. Our sensors can be embedded inside the reinforcement during weaving and they have all the characteristics of a traditional textile material (flexible, lightweight and are capable of adopting the geometry of the reinforcement and become its integral part).

Embedding such an intelligent piezoresistive sensor inside the reinforcement during weaving process is the most convenient and cost effective way of insertion of a sensor for structural health monitoring.

Development and optimization of such piezoresistive sensors has been carried out in order to render them sensitive enough to measure in situ strains inside the composite part. Sensitivity is important as the targeted application usually undergoes very low strains and even such low strains and/or vibrations during the life time of composite parts are critical. Often they are used in areas where structural integrity can not be compromised (aircraft wings, bodies etc).

2.1. Sensor design and optimisation

As coating solution, the conductive polymer composite based on dispersion of carbon black particles (Printex® L6) in polymer (Evoprene® 007) solution, using chloroform as a solvent was chosen (Cochrane et al, 2007, 2010).

In order to characterize the sensitivity and adherence of the coating on different substrates, the 35 % carbon black solution was coated on different yarns (71 tex cotton spun yarns; 482.3 tex polyethylene monofilament and 25 tex polyamide monofilament). Visual inspection of the surfaces of the coated yarns shows that the coating is more uniform for synthetic monofilaments compared to cotton yarns. The cotton yarns absorb the conductive solution, which penetrates inside the pores and interstices much like a dye. This particular phenomenon could be a source of non homogeneity in sensor electrical and mechanical properties, as the spun yarn is non uniform as compared to filaments, the coating and thus the resistivity achieved could be non uniform. Moreover the greatest inconvenience with coated cotton spun yarns is their low sensitivity during the initial tensile loading phase.

The resistance values were measured on 12 coated samples of each variant using a multimeter. The resistivities were then calculated using the yarn/filament fineness, yarn/filament lengths and these measured resistance values. Fig. 1 gives a comparison of calculated resistivity values of conductive layer deposited on different fibrous substrates. It can be seen that coated polyethylene filaments show relatively lower dispersion of resistivity as compared to coated poyamide filaments.

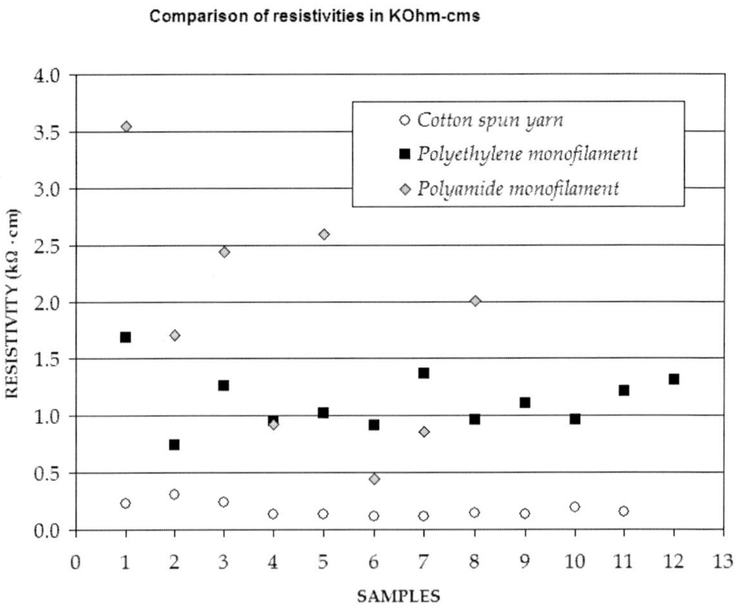

Figure 1. Resistivity values calculated for different substrates coated with 35 wt.- % carbon black solution

In order to carry out tensile tests on coated yarns and monofilaments, MTS 1/2 tester was used. Samples underwent quasi static tensile loading at a constant test speed of 5 mm/min. For the purpose of electrical resistance variation measurement during the tensile testing, a simple voltage divider circuit and Keithley KUSB-3100 data acquisition module were employed. Fig. 2 shows some of the results for electrical resistance variation, expressed as normalised resistance ($\Delta R/R$) during tensile testing, obtained using different substrates for coating.

Initial resistance of the coating on cotton yarns is much lower then monofilaments. But since the cotton spun yarns are inherently irregular, the coatings obtained are not homogenous and the results for different coated yarns vary widely in their response to tensile loading (Fig. 2-a). Due to particular fineness of the polyamide monofilament it was found that slight non homogeneity in coating on the surface can result in breakdown of conductive path as is obvious from Fig. 2-b. As a result, the behaviour of polyamide is highly inconsistent. Polyethylene monofilaments provide a reasonably good compromise as the substrate. The coatings on polyethylene are easy to achieve due to good substrate/conductive solution interfacial properties. As the curves in Fig. 2-c show, the polyethylene coatings are reproducible as the curves for all the four samples are nearly identical as opposed to polyamide and cotton. Therefore, polyethylene monofilament was chosen for sensor development.

The two ends of the coated polyethylene filaments were additionally coated with silver paint and fine copper wire was attached to the two ends with the help of this paint (as shown in the Fig. 3). In this way, secure connections were realized enabling the reduction of the contact resistance to the minimum.

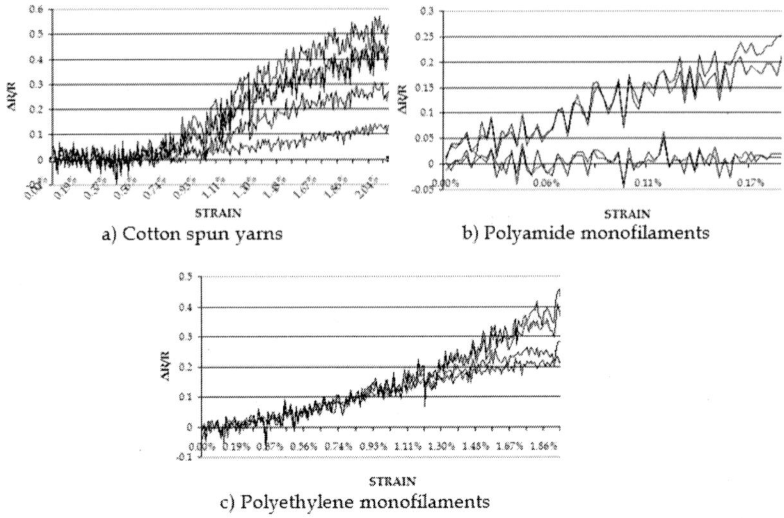

Figure 2. Electrical resistance variation during tensile strength tests on different yarn and filament substrates coated with 35 wt.-% carbon black solution

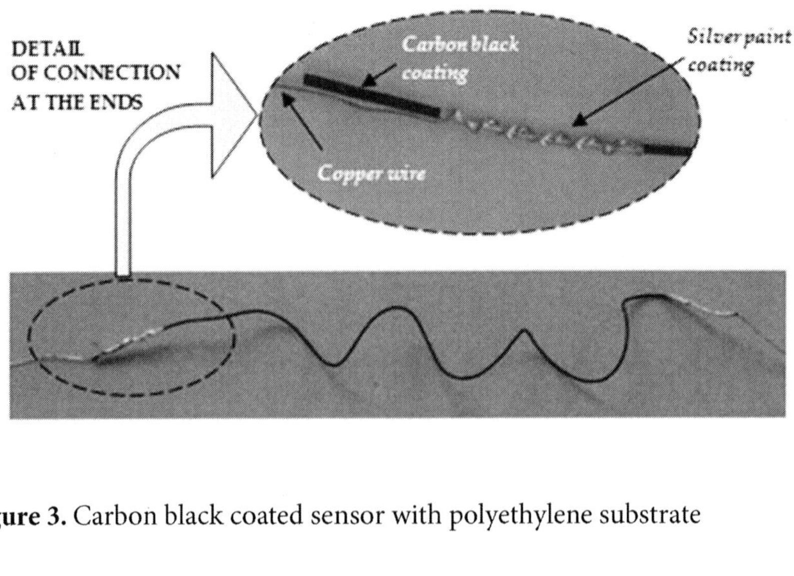

Figure 3. Carbon black coated sensor with polyethylene substrate

Sensor structural and geometrical parameters along with initial electrical resistance are shown in Table 1.

Table 1. Sensor properties

No.	PARAMATER	UM	VALUE
1	Linear density of the filament	g/km	48.23
2	Diameter of the filament	mm	0.70
3	Average width of the sensor cross section	mm	1.68
4	Average thickness of the sensor cross section	mm	1.26
5	Aspect ratio of the sensor (width/thickness)	-	1.33
6	Initial resistance of the sensor	kΩ	43.3

For insertion in conductive fibre based reinforcements like that woven using carbon multifilament tows, the sensor was coated with Latex Abformmasse supplied by VossChemie® so as to insulate the sensor from surrounding carbon tows.

Prepared in this way, the sensor with polyethylene substrate was tested again on MTS 1/2 tester, under quasi static tensile loading at a constant test speed of 5 mm/min. The same Keithley® KUSB-3100 data acquisition module was employed for the purpose of voltage variation during tensile testing. This time, a special set-up containing a Wheatstone bridge and an amplifier was used to measure unknown variable resistance of the sensor as a function of output voltage. As is obvious from curves presented in Fig. 2-a, b and c, the simple voltage divider circuit is not adequate for the measurement of resistance change in case of sensors developed here. These piezoresistive sensors produce a very small percentage change in resistance in response to physical phenomena such as strain. Moreover the output signal has considerable noise. Generally, a bridge measures resistance indirectly by comparison with a similar resistance. Wheatstone bridges offer an attractive solution for sensor applications as they are capable of measuring small resistance changes accurately (Wilson, 2004).

Fig. 4 shows schematic diagram of the data acquisition module developed and used for data acquisition and its further treatment.

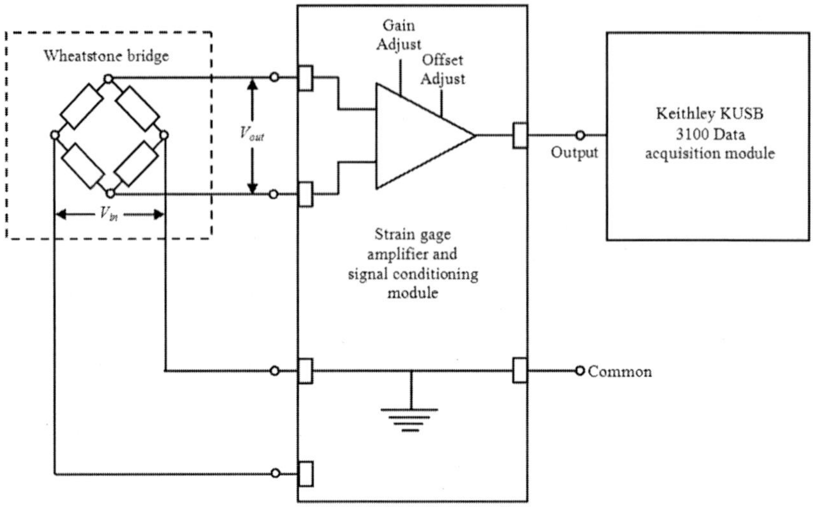

Figure 4. Schematic of instrumentation amplifier (INA 101) connected to Wheatstone bridge

The resistance variation data thus obtained for different test results was treated for noise reduction using a low pass filter. The resultant stress-strain-resistance relationship curve up to 2.75 % elongation of the out of composite sensor (before insertion in the reinforcement) is shown in Fig. 5.

It may be noticed in Fig. 5 that the stress vs. strain curve has the same shape as normalised resistance (ΔR/R) vs. strain curve. This validates electromechanical properties of our fibrous sensor for strains ranging from 0 to 2.75 %.

In Fig. 6, the hysteresis results of the sensor for 10 cycles have been given. The sensor underwent 0.5 % extension at a constant test speed of 5 mm/min, followed by compression in each cycle. The sensor follows the extension and compression patterns in each cycle.

The hysteresis is high for the first cycle which reduces gradually and for the 10th cycle the sensor exhibits almost linear behaviour.

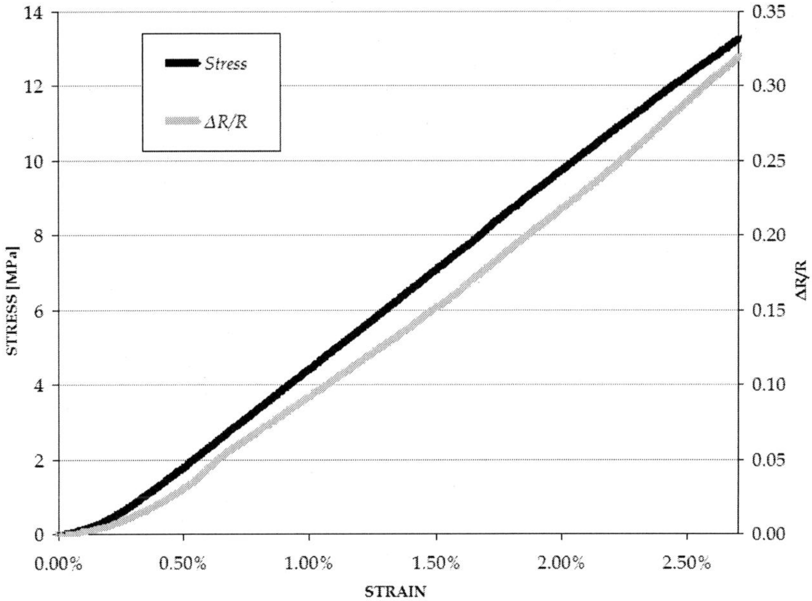

Figure 5. Normalized resistance and stress against strain for sensor outside composite

2.2. Sensor insertion in carbon woven reinforcement

An orthogonal/layer to layer warp interlock with 13 weft layers and 12 warp layers was chosen as woven structure (Fig. 7-a) and than was woven on a modified conventional loom (Patronic B60 ARM). 6K multifilament carbon tows (supplied by Hercules Inc.) having 200 tex was used in both direction – warp and weft. Yarn densities were 24 yarns/cm in warp direction and 170 yarns/cm in weft direction. The thickness and areal density of resulted reinforcement were 6.5 mm and 3908 g/m^2, respectively.

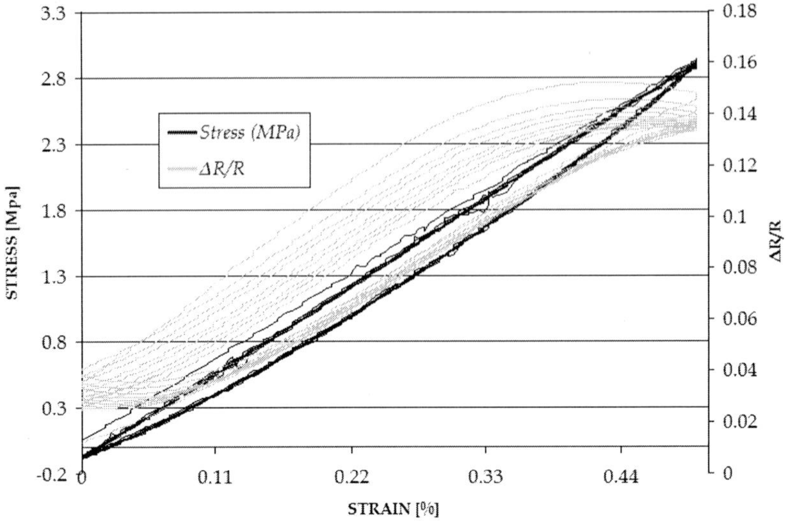

Figure 6. Normalized resistance (ΔR/R) and stress against strain for sensor (Hysteresis 10 cycles at 0.5 % extension)

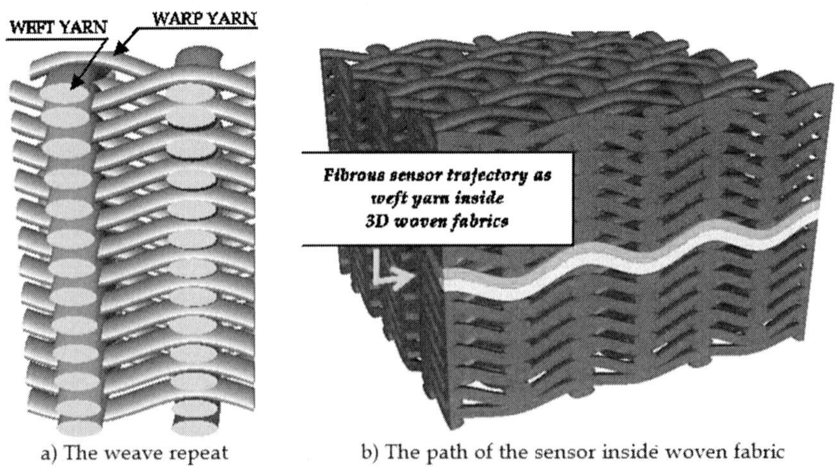

Figure 7. Interlock weave structure used as reinforcement – graphical representation (TexGen software)

Sensors can be inserted in warp or weft directions during weaving. Given the technical complications associated with sensor insertion in warp direction during weaving on a loom, insertion in weft direction has been carried out for preliminary studies. The placement of sensor in the reinforcement was decided so that the sensor was inserted in the middle of the structure related to thickness (Fig. 7-b).

The sensor was inserted during the weaving process, as a weft yarn and it follows the same trajectory as the carbon weft yarns inside the reinforcement. In Fig. 8, off the loom dry reinforcement photograph have been shown. Latex coated sensor connections can be seen protruding from the reinforcement.

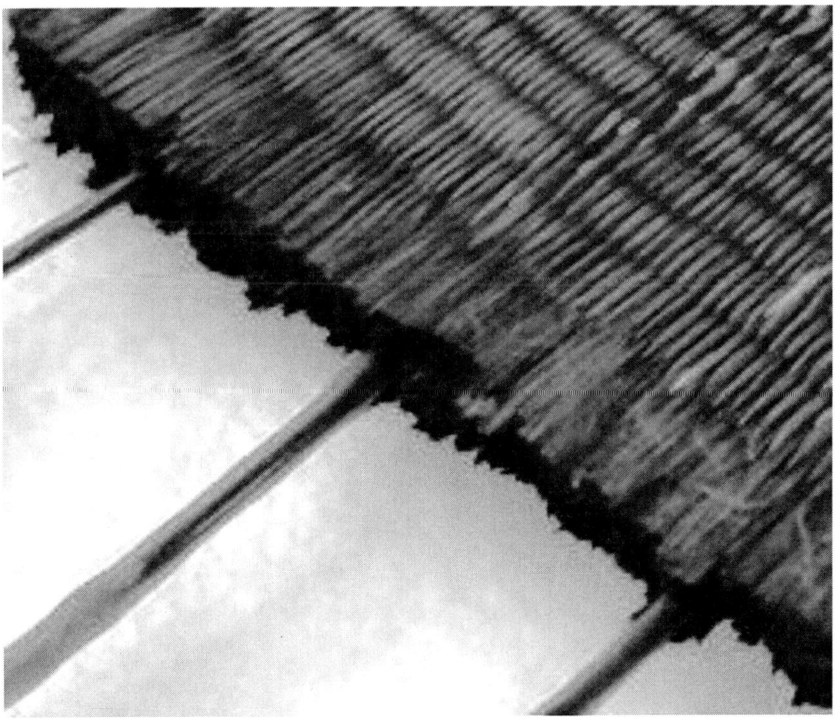

Figure 8. Reinforcement with protruding sensor connections

2.3. Carbon woven reinforcement impregnation and testing

After weaving, the reinforcement was carefully removed from the loom and was impregnated using vacuum bag infusion process in order to make the composite part stiff. The resin employed was epoxy Epolam® 5015. The two connections of the sensor which remain outside the reinforcement at the two ends were carefully separated from the rest of the mould. This was done by creating two vacuum sub moulds inside the larger mould so that the resin may not impregnate the two connections of the sensor. The impregnated composite samples were cut into slabs of 25 X 2.5 cm (Fig. 9).

The composite specimens were tested on Instron 8500 tester. Tensile strength tests were performed on the composite specimens (according to ISO 527-4, 1997) in the weft direction i.e., the direction parallel to the inserted sensor. The same Wheatstone bridge was used for resistance variation measurement. The configuration of the testing equipment was also kept the same. The composite structural part was tested at constant test speed of 5 mm/min. The composite underwent traction until rupture.

2.4. On-line measurements of sensor in woven fabric composites - Results

Resultant stress-strain-resistance relationship curve is shown in Fig. 10. It can be observed that the normalized resistance follows the stress-strain curve. The stress-strain-resistance curve can be divided into four regions: *the initial stiff region* - where the composite exhibits toughness against the applied load represented by high slope; *the tows straightening region; the second stiff region* and the *zone of rupture*. The rupture occurred at the strain of 0.52 %, after which the tensile strength tester came back to its initial position at the same speed (5 mm/min). Since the fibrous sensor has not been broken, the normalised resistance ($\Delta R/R$) decreased until zero as the tester returned to its initial position. However this decrease was not linear because the sensor was still intact while the resin-sensor interface was partially damaged which caused its non linear behaviour.

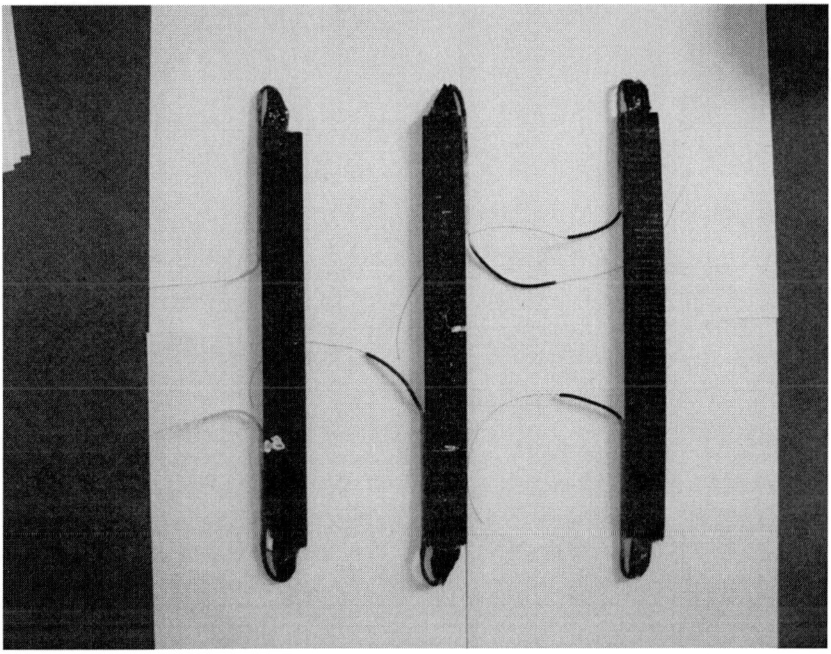

Figure 9. Textile composite sample containing fibrous piezo-resistive sensor

Due to the high difference in yarn densities (24 warp yarns/cm vs. 170 weft yarn/cm), the weft tows are highly crimped. In the initial stiff region micro-cracks start appearing as the composite specimen undergoes traction but the interface at resin and multifilament tows is still intact. That is why the composite exhibits rigid behaviour. In Fig. 10 it can be observed that after the initial stiff region the highly crimped tows tend to straighten due to increasing tensile load in the second region. In this region the micro-cracks give way to relative slippage of highly crimped tows in their sockets i.e., the resin-tow interface is relatively weakened. It can also be remarked that the sensor resistance follows the stress strain curve, but in the second region the electrical resistance curve is noisier as compared to other regions of the curve which might signify the slippage of tows as well as the sensor in their sockets. This second region is followed by the third region called the second stiff region where the tows are locked in their sockets. In this region the tows resist the applied load and exhibit stiff behaviour as they regain some of their initial stiffness after the straightening of tows in the second region. The electrical resistance varies almost linearly with the applied load, in this region. The third region is followed by the zone of rupture of the composite in which the electrical resistance, having attained the highest value starts dropping down. The normalized resistance starts dropping after the rupture. The fact that the sensor resistance attains its initial value after the rupture signifies that the sensor, owing to its elastic properties, is not destroyed with the composite. This fact was confirmed by tomographical image of the samples which underwent traction, shown in Fig. 11-a) and b). Sensor cross section and its path at and near the zone of rupture can be observed.

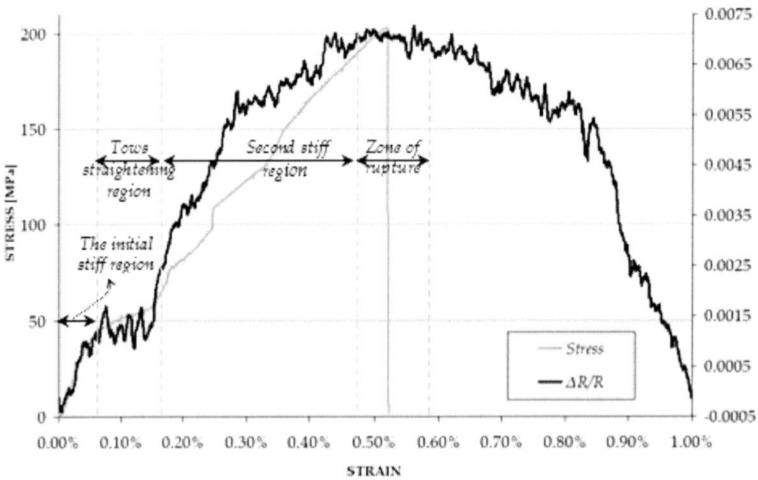

Figure 10. Normalized resistance and stress against strain for sensor inside composite

In Fig. 11-a) and b) it can be observed that the sensor-resin interface has a lot of voids. These are caused by poor resin-sensor interfacial properties. The insulating medium on the sensor surface needs to have good adherence with the epoxy resin and carbon fibre reinforcements. Damage that occurred at the main rupture zone has propagated along the sensor boundary giving rise to de-bonding of the sensor. A kink in the sensor can be observed which is caused by the relaxation of sensor as it tries to regain its original dimensions after the tensile loading damages the composite sample. The insulation coating around the sensor renders it thick as well which is undesirable for high performance composite materials as thick insulation coatings might adversely affect the mechanical properties.

3. HEATING ELEMENTS BASED ON CONDUCTIVE POLYMER COMPOSITE

The second part of this chapter presents a woven fabric containing an original heating element. Textile actuators like heating fabrics can find applications in numerous and varied fields such as sports, leisure, medical and automotive (Droval et al., 2005; El-Tantawy et al., 2002). In garments, wearability is affected because of the use of metallic components (heating wire and/or heating track on polymer flexible substrate), that are rarely elastic, flexible and lightweight. Nevertheless, these metallic, non-textile elements can be replaced by other conductive fibres such as silver plated polyamide fibres. In that case, the heating textile becomes lightweight, but very expensive (WarmX GmbH). In all the cases, heating systems need heavy power supplies. Thus, it is very important to develop heating textile systems able to work at low voltage.

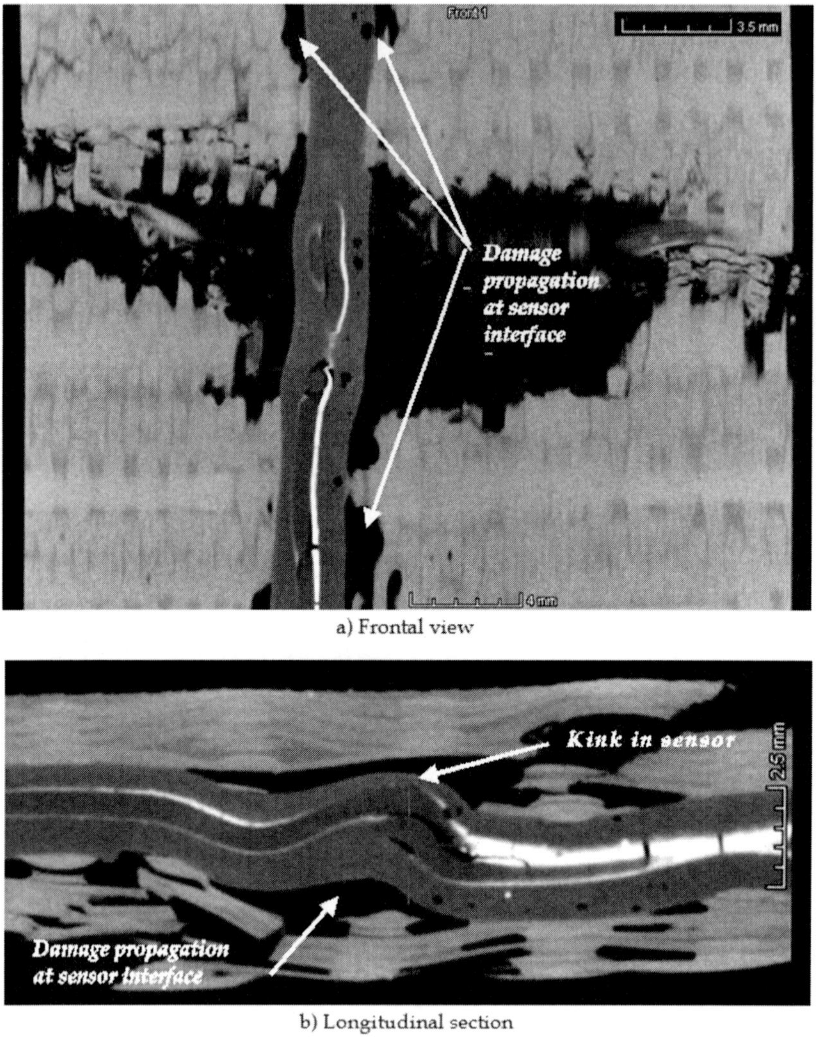

Figure 11. Tomographical images of sensor inside a tested sample near the zone of rupture

Our heating element is designed to adapt to woven flexible structures. Additional metallic yarns, used as electrodes, are integrated in a woven structure (or sewn into textile) in a comb-teeth arrangement. Function of these electrodes is to connect heating textile to a power supply and to distribute the current in the conductive coating layer applied on the fabric surface. The comb-teeth electrode arrangement is specially designed to ensure uniform heat distribution. The coating is realized with a composite material based on aqueous latex dispersed with carbon black (CB) as filler. The heating element (comb electrodes and electro-conductive coating) can thus adopt the desired pattern. This is an important aspect of our heating element as it allows integration of the heating element in various fabrics designed for varied and diverse applications.

3.1. Materials and methods

Comb structure was made with stainless steel yarns (2 x 275 x 12 μm from Bekintex®). The average yarn count was 500 Tex, with a resistivity of 14 ohm/m. These yarns were either woven or sewn on an existing fabric. The common feature of all the configurations is that only one comb-teeth structure was used (Fig. 12). The textile fabric was woven on a hand loom (ARM loom equipped with Selectron command box). A plain weave was chosen. Cotton yarns were used in warp and weft having densities of 27 and 10 yarns/cm respectively. The stainless steel yarns were introduced manually during the weaving process according to the pattern (Fig. 12).

Samples with heating surface (*i.e.* L x l in Fig. 12) larger than 180 cm^2 were prepared. In typical samples, the dimension *L* was about 140 mm while *l* was about 150 mm. In this study, the distance between electrodes (*lp*) remained unchanged: i.e., 20 mm.

The coating was made using a conductive polymer composite (CPC) composed of carbon black (CB, Printex ® L6, Degussa), a synthetic rubber latex solution (Kraton® IR-401, Kraton Polymers) a dispersing agent (Disperbyk®-2010, SPCI) and water.

The preparation procedure is as follows: the dispersing agent is put into water and the CB particles are gradually added while mixing continuously. The polymer is finally added while mixing gently in order to avoid too strong shearing. The coating was then applied on the fabric with a magnetic coating table equipped with a magnetic bar as scraper. 12 samples were prepared with different CB content: 2.5, 5.0, 7.5, 10.0, 15.0, 20.0, 30.0, 35.0, 40.0, 45.0, 50.0 and 60.0 wt.-%. These contents were calculated from the total weight CB + Latex solution. After coating the woven fabric samples were dried at 50 °C for 12 hours. For all the samples, the thickness of the final coating layer was 450 ± 50 μm.

For each coating surface resistivity was measured using four-point probe (MR-1 Surface resistance meter, Schuetz Messtechnik). The aim of these measurements is to determine the percolation threshold and the minimum CB content which allows sufficient electrical conduction for our application.

To characterize heating effect of the samples, 2 processes were used:

- Feeding of the heating element with variable voltage supply (10, 15, 20 and 24 V). The surface temperature was recorded using thermocouple at 15 minute intervals at 5 different locations of the fabric. The average temperature was calculated from 5 measurements. Ammeter was used to determine the power consumption (W) of the heating element. This consumption is expressed in mW/cm^2 (taking into consideration the surface area of each sample),
- Feeding of the heating element with constant voltage supply (15 V) in conjunction with an IR camera (Agema ThermoVision 900). This camera took an IR image every 20 seconds.

3.2. Results

Fig. 13 shows electrical resistivity of coatings plotted against filler (CB) content in the latex solution. As expected, it is possible to identify the percolation threshold from this plot, which lies at 12 ±1 wt.-%. The form of the plot is in accordance with the typical behaviour of systems consisting of percolation networks (Kirkpatrick, 1973).

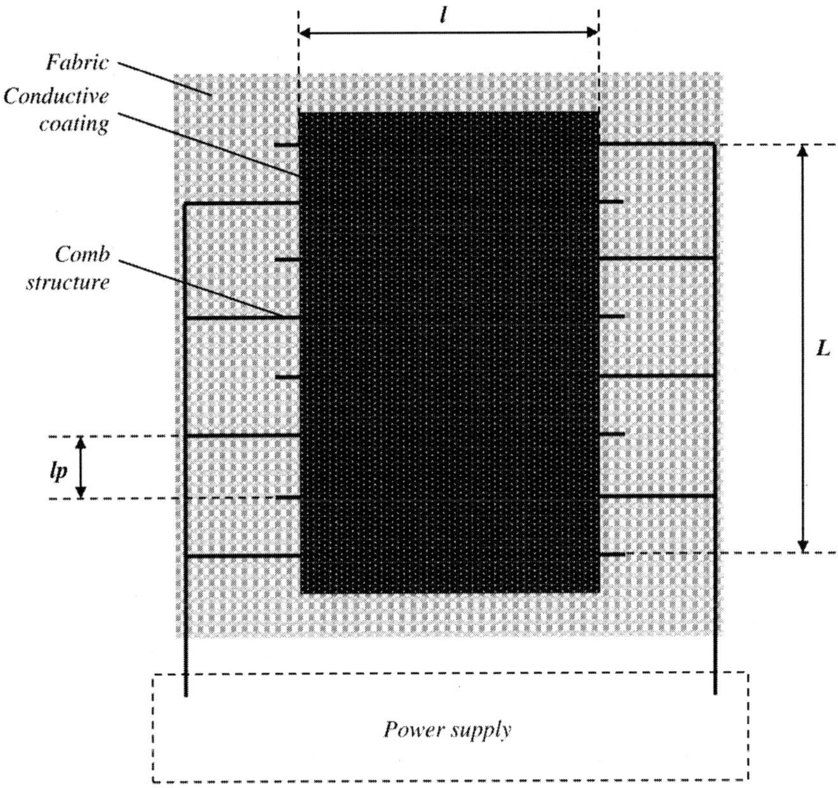

Figure 12. General structure (comb-teeth pattern and conductive coating) of heating textile element

This threshold value in wt.-% is expressed for liquid latex solution. Liquid latex contains approximately 63 % of dry material by weight. Thus, the corrected value of percolation threshold is near 18 wt.-%. This value is relatively higher than the value reported in literature for similar systems, (Grunlan et al. 1999, 2001). In our study, process of dispersion (including rupture of CB aggregate) and coating on fabric is not yet optimized. Obtained results show that 15 wt.-% of CB is necessary to obtain à conductive coating. Nevertheless, Fig. 13 shows that between 15 and 40 wt.-% resistivity is not optimal: therefore it is necessary to fill the composite at least by 45 wt.-% to obtain lower resistivity.

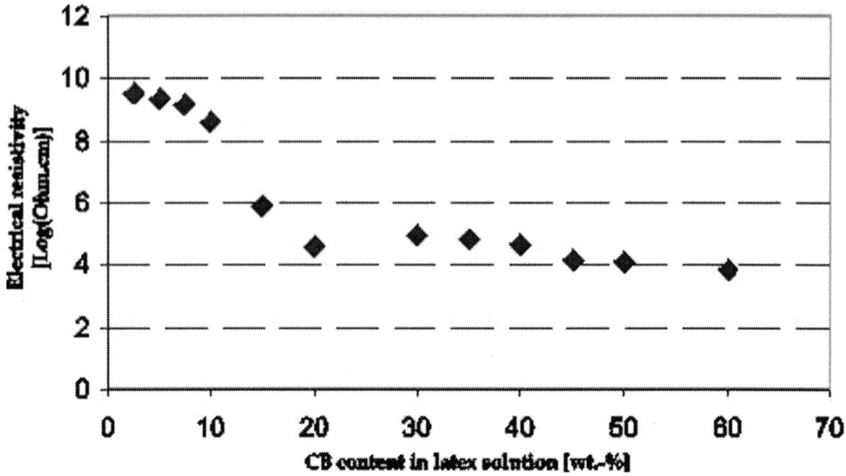

Figure 13. Electrical resistivity of the coating vs. CB content in the latex solution

Fig. 14 shows surface temperature of coating plotted against the filler content (in liquid latex) for several feed voltages (10, 15, 20 and 24 V). Temperature on the graph (ΔT) is expressed as difference between measured temperature and room temperature (between 20 and 22 °C). No elevation of temperature was recorded for sample under 30 wt.-% of CB. For CB content between 30 wt.-% and 45 wt.-%, ΔT increases with the CB content. Above 45 wt.-% of CB, ΔT does not increase significantly with filler addition. These results are in agreement with the previous remarks concerning resistivity vs. CB content.

These results show that the best content of CB was, in our case, 45 wt.-%. Under this value heating effect was non optimal. Above this value addition of CB does not increase heating ability.

Maximum value of ΔT (near 20 K) was registered for sample voltages of 24 V and CB content of 45 wt.-% and 60 wt.-%. Fig. 15 shows that for these heating elements, electrical input power was close to 250 mW/cm^2.

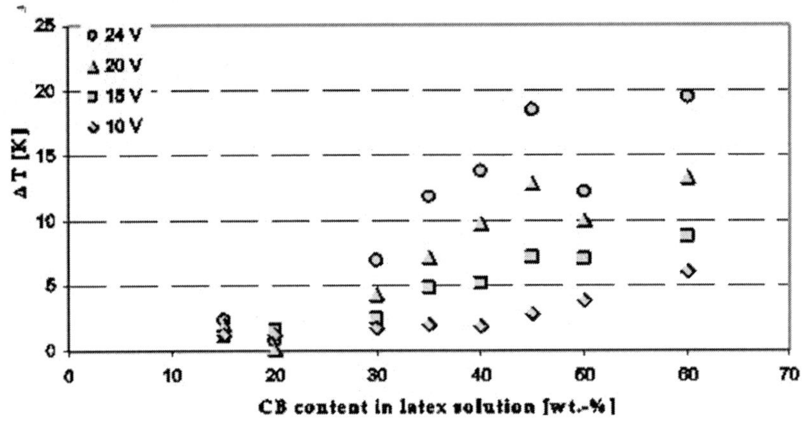

Figure 14. Surface temperature vs. CB content for feeding voltage of 10, 15, 20 and 24 V

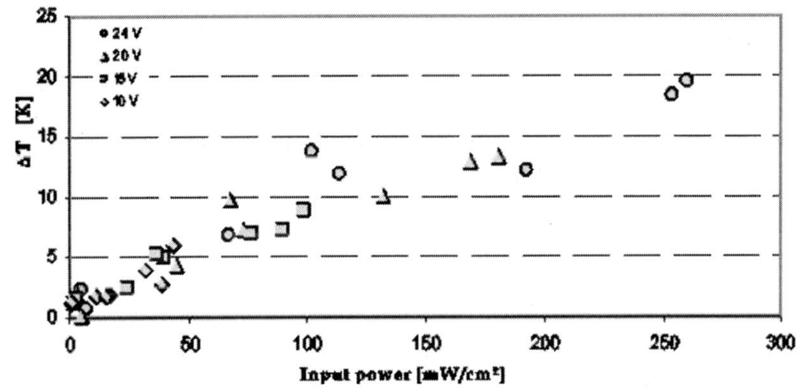

Figure 15. Increase of surface temperature (ΔT) vs. power consumption of heating element

Infrared images show the distribution of heat in the structure (comb electrodes and conductive coating) vs. time. Fig. 16 shows temperature of sample with 60 wt.-% of CB from t = 0 s (Fig. 16-a) to t = 180 s (Fig. 16-f). Feeding voltage was constant and was equal to 15 V.

Fig. 16 shows that comb structure (stainless steel yarn) heats first after switching on. The maximum temperature of this yarn was about 70 °C.

This temperature is achieved after 120 s. CB coating heats relatively slowly but it can be conjectured that after 120 s the temperature of all the surface area exceeds 35 °C while it exceeds 40 °C after 180s. This behaviour is expected since stainless steel yarns have better thermal conductivity than carbon based composite.

4. CONCLUSION

The sensor based on conductive polymer composite, developed for in situ measurements on carbon fibre based woven fabric composite, is capable of detecting strain in the structure. The electrical resistance variation in the sensor follows the deformation pattern of the composite, mainly due to its sensitivity to its environment and because of the fact that it is integrated in the structure and follows the fibre architecture of the reinforcement. It has been shown that the integrated textile sensors inside the reinforcement can be used as in situ strain gages for the composite materials. Moreover, if the placement of these sensors inside the reinforcement is carefully chosen, they can be used to follow the local deformation pattern so as to better understand the deformation mechanisms and predict life time of the composite parts. At present the sensors have been tested for tensile loading. Tensile strength tests were chosen to demonstrate the basic features of this novel SHM approach. In the future these sensors will be used for bending and fatigue tests on similar 3D carbon fibre woven reinforcement based composites. However optimisation of sensors needs to carried out in order to prepare finer sensors having negligible affect on reinforcement geometrical and mechanical properties. For carbon fibre based reinforcements which require an insulation coating on the sensor surface, a better and finer coating needs to be applied. In view of the test results presented, it can be concluded that these sensors can be used for in situ health monitoring of various types of composites for industrial applications (aeronautics, automotive etc).

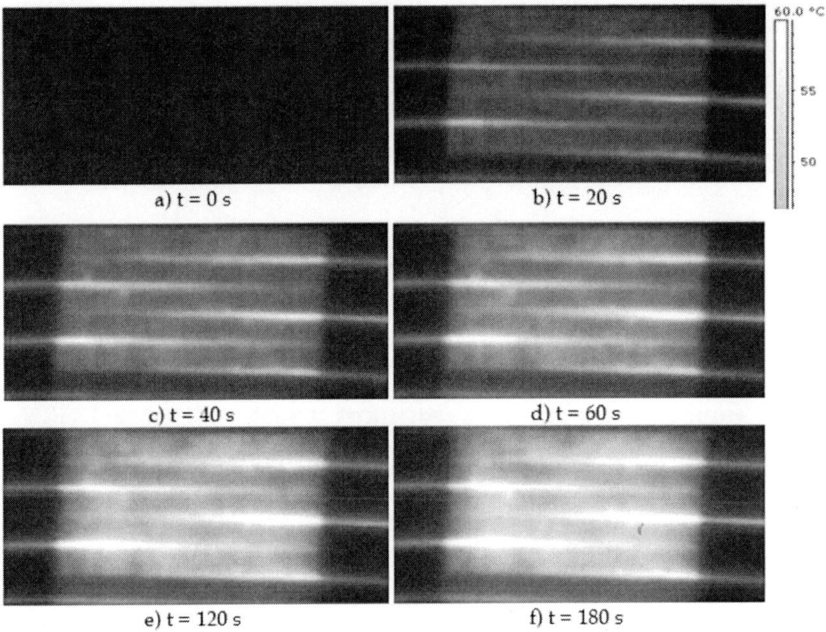

Figure 16. IR image of 60 wt.-% CB heating element from a) t = 0 s to f) t = 180 s, voltage – 15V

We have also developed a heating element based on original comb-teeth structure (stainless steel electrode) and electro-conductive coating composed of latex and carbon black. Comb structure can be either woven or sewn into the fabric. Final product is flexible and lightweight. This study has show that ideal carbon content of the coating was 45 wt.-% in a latex solution. Our heating elements (about 200 cm^2) allow increasing the temperature (ΔT) by 20 K with low voltages (between 15 and 24 V). Temperature homogeneity of heating elements is better than those of heating elements made with only stainless steel yarns because in the later case, heat is only produced and localized at conductive yarns. Moreover, our system is cost efficient because we used a few stainless steel yarns.

The next step of this study would allow, in the first place, the optimization of composite preparation (liquid Latex + CB) and secondly the realization of larger heating elements (> 0.5 m^2). Potential applications of these heating elements can be found in garments (to improve thermal comfort), in transportation where heating is required for passenger comfort and in certain industrial systems (antifreeze).

5. ACKNOWLEDGEMENTS

This research was partially financed by Interreg IV via TRITEX (Transfer of Research and Innovations in TEXtiles) program (FLV1.1.1). Special acknowledgments to Textile Department Laboratory from UGENT (Belgium) for electrical measurements (part 3.) and use of IR camera.

REFERENCES

1. S. Black, 2009 Structural health monitoring: Composites get smart, Composites World, http://www.compositesworld. com/articles/ structural-health-monitoring-composites-get-smart

2. C. Cochrane, V. Koncar, M. Lewandowski, C. Dufour, 2007 Design and development of a flexible strain sensor for textile structures based on a conductive polymer composite. Sensors, 7 473492 .

3. C. Cochrane, M. Lewandowski, V. Koncar, 2010 A Flexible Strain Sensor Based on a Conductive Polymer Composite for in situ Measurement of Parachute Canopy Deformation. Sensors, 10 82918303 .

4. P. Dharap, L. Zhiling, S. Nagarajaiah, E. V. Barrera, 2004 Nanotube film based on single-wall carbon nanotubes for strain sensing, Nanotechnology, 15 379382 .

5. G. Droval, P. Glouannec, J. F. Feller, P. Salagnac, 2005 Simulation of electrical and thermal behavior of conductive polymer composites heating elements. Journal of Thermophysics, 19 3 375381 .

6. F. El -Tantawy, K. Kamada, H. Ohnabe, 2002 In situ network structure, electrical and thermal properties of conductive epoxy resin-carbon black composites for electrical heater applications. Materials Letters, 56 112126 .

7. European Standard NF EN ISO5274 1997 (), Plastiques-Détermination des propriétés en traction, Partie 4: Conditions d'essai pour les composites plastiques renforcés de fibres isotropes et orthotropes.

8. B. Fiedler, F. H. Gojny, M. H. G. Wichmann, W. Bauhofer, K. Schulte, 2004 Can carbon nanotubes be used to sense damage in composites?, Annales de chimie, 29, 8194 .

9. C. L. Gettinger, A. J. Heeger, D. J. Pine, Y. Cao, 1995 Solution characterization of surfactant solubilized polyaniline. Synth. Met., 74, 8188 .

10. J. Grunlan, W. Gerberich, L. Francis, 2001 Lowering the percolation threshold of conductive composites using particulate polymer microstructure. Journal of applied polymer science, 80 4 692705 .

11. J. Grunlan, W. Gerberich, L. Francis, 1999 Electrical and mechanical property transitions in carbon-filled poly(vinylpyrrolidone). Journal of materials research, 14 11 41324135 .

12. Y. Haba, E. Segal, M. Narkis, G. I. Titelman, A. Siegmann, 1999 Polymerization of aniline in the presence of DBSA in an aqueous dispersion. Synth. Met., 106 5966 .

13. A. J. Heeger, 2002 Semiconducting and metallic polymers: the fourth generation of polymeric materials. Synth. Met. 125 2342 .

14. R. Kamiya, B. A. Cheeseman, P. Popper, T. Chou, W. , 2000 Some recent advances in the fabrication and design of three-dimensional textile preforms: a review. Composites Science and Technology, 60 3347 .

15. S. Kirkpatrick, 1973 Percolation and conduction. Reviews of Modern Physics, 45 4 574588 .

16. F. Ko, 2007 3 -D textile reinforcements in composite materials, 3-D Textile Reinforcements In Composite Materials. vol. null: Woodhead Publishing Limited.

17. I. Krupa, I. Chodak, 2001 Physical properties of thermoplastic/ graphite composites. Europ Polym J., 37, 21592168 .

18. D. Kumar, R. C. Sharma, 1998 Advances in conductive polymers. Eur. Polym. J. 34 10531060 .

19. C. Li, E. T. Thostenson, T. W. Chou, 2008 Sensors and actuators based on carbon nanotubes and their composites: A review, Composites Science and Technology, 68 12271249 .

20. F. Lorussi, E. P. Scilingo, M. Tesconi, A. Tognetti, D. De Dossi, 2005 Strain sensing fabric for hand posture and gesture monitoring, IEEE Transactions On Information Technology In Biomedicine, 9 372381 .

21. I. Novak, I. Krupa, I. Chodak, 2002 Investigation of the correlation between electrical conductivity and elongation at break in polyurethane-based adhesives. Synth. Met., 131 9398 .

22. E. P. Scilingo, F. Lorussi, A. Mazzoldi, D. De Rossi, 2003 Strain-sensing fabrics for wearable kinaesthetic-like systems, IEEE sensors journal, 3 460467 .

23. J. Wilson, 2004 Sensor signal conditioning, Sensor Technology Handbook. vol. null: Newnes Publishing Limited.

24. www.nottingham.ac.uk/~emxmns/texgen.htm, consulted on 10/01/2011

25. www.datasheetdir.com/INA101 -Instrumentation-Amplifiers, consulted on 06/05/2010.

CHAPTER 3

Smart Woven Fabrics in Renewable Energy Generation

Derman Vatansever[1], Elias Siores[1], Ravi L. Hadimani [1] and Tahir Shah[1]

[1] *University of Bolton, Institute for Materials Research and Innovation, United Kingdom*

1. INTRODUCTION

Initially, the first purpose of fabric making was only for covering the body and sheltering. However, with a growing population and ever improving advanced technologies, today's fabrics are mostly used for fashion and performance thus enhancing the standard of people's everyday life and enjoyment.

Most of the technologies which increase the standard of living also increase carbon emission and adversely affect human life indirectly. Warming buildings, using cars, provide hot water, cooking food etc. need energy generated by using coal, gas, fuel or electricity. Burning gas and fuel accelerate the threat of nature and eventually contribute to global warming. Even electricity generation causes carbon emission unless it is generated by using renewable energy sources. Interest in providing renewable usable electrical power from the environment has grown, particularly in the elimination of battery usage, because of their sizeable dimensions, weight and limited lifetime.

Since global warming is being considered as the biggest danger for the nature, many scientists and researchers have brought a new breath to their researches. As almost all areas of renewable science and technology, researchers are now working in the field of textile fabrics capable of generating green electricity.

Undoubtedly, weaving is the oldest fabric making method which has been a part of human life for protection from nature's elements and hazards. It is now possible to produce smart woven fabrics by combining the oldest fabric making method with smart fibre material technologies. The chapter named "Smart Woven Fabrics in Renewable Energy Generation" contains a brief introduction to smart materials, focusing on piezoelectricity and polymer based piezoelectric fibre production. The rest of the chapter explains how to produce smart woven structures by integrating smart fibres into the fabric during weaving process and examples for possible applications for energy regeneration from nature's elements are given.

2. BRIEF INTRODUCTION TO WEAVING AND LOOMS

2.1. Vertical and horizontal looms

The first loom consisted of only a branch of a tree that is parallel to the ground. In this simple design, warp threads were directly fastened to the branch of a tree and held parallel to each other under tension caused by tied stones at the other end of the warps. The weft threads work from right to left and left to right by passing through hanging warps until the designed fabric is composed. Later, a simple framework was made of tree branches to form a vertical loom to use for indoor as well as outdoor.

(a) (b)

Figure 1. a) A vertical warp-weighted loom used in Iceland, reconstructed and set up before 1914, National Museum of Iceland-Photograph by Gilsli Gestsson (**Broudy, 1979**),(b) a horizontal ground loom used in Northen Cameroon-Photograpgh by René Gardi (**Broudy, 1979**).

Figure 1 (a and b) show examples for early vertical loom and horizontal looms and Figure 1 shows a laboratory scale hand loom (c) and automated loom (d). Warp threads are attached to heddles and held parallel to each other on the loom and weft thread is held by a shuttle. The heddles keep warp threads apart so that the shuttle could pass through easily and cary the thread from left to right and right to left so that the weft will work over and under warp threads to form a fabric. Movements of the heddles on hand looms are controlled by pedals beneath the loom.

Current generation looms (Fig. 1d) are fully automated. Movements of the heddles are automatically controlled according to the fabric design sent via a computing system. The important parameters, such as tension, speed, temperature etc., are measured and/or controlled by the controlling systems. Thus, any errors that occur during the fabric production are immediately detected.

(c) (d)

Figure 1. c) An example of laboratory scale hand looms and (d) an example of automated looms (**wikipedia, 2011**).

2.2. Main weaving constructions

Weaving is one of the traditional fabric making methods. There are two sets of threads, warp threads and weft threads which form a fabric by being interlaced row by row. Weaving has three main construction techniques; plain weaving, twill weaving and satin weaving. Any other techniques developed are variations of these main techniques.

Plain weave is created by interlacing the weft across the warp threads and for this at least 2 heddles are needed. Weft thread goes under a warp thread and then over the next one so that the equal amount of weft and warp is seen on both surfaces of the woven fabric.

Twill weave is created by interlacing two or more weft threads over and under one or more warp threads so that at least three heedles are needed to make a twill woven fabric. In twill wovens, parallel diagonal ribs are formed from left to right or from right to left that depends on the formation, woven fabric is characterised and named as “S“ or “Z“ twill, respectively. If a twill woven has more warps than the wefts on the fabric face is known as warp faced twill and if it has more wefts on the fabric face, it is known as weft faced twill.

Satin weave is created by floating four or more yarns before a single interlacing occurs and at least 5 heddles are needed to make a satin woven fabric. The warp or weft threads pass across many threads in such a way, like 4/1, 5/1, 7/1 etc., that the face of the woven fabric is mostly covered with either warp or weft threads and known as warp satin or weft satin, respectively.

Depending on the construction of the weave and characteristics of the used warp and weft threads, fabric properties may vary. Count of the used fibres/yarns determines the fabric density which affects the weight, feel, appearance, thickness etc. Variable properties are many and can be found in the literature (Gioello, 1982). They include:

- Weight
- Hand and feel
- Drapeability
- Appearance
- Covering power
- Surface texture
- Body fit

- Thickness
- Lustre
- Strength
- Flexibility / pliability
- Resiliency
- Warmth
- Affinity to dyestuff

Fabrics created by using plain weave technique are tough compared to twill and satin weave. The fabrics made by twill weaving technique are more pliable and drapable than plain weave fabrics but not as pliable as satin weave. However, strength of plain woven fabrics is higher than both twill and satin woven fabrics.

3. SMART MATERIALS AND PIEZOELECTRICITY

A smart material is one that shows extraordinary response when subjected to a stimulus. Piezoelectric materials are considered as smart materials because of their ability to generate electricity against the stimulus which is mechanical strain or vibration (Fig.2a). This property of piezoelectric materials is known as direct piezoelectric effect. The reverse effect is also possible in that this material undergoes a slight deformation in shape when a small electrical field is applied (Fig.2b).

Direct piezoelectric effect of the materials is also known as "generator effect" or "sensor effect". The terms "motor effect" and "actuator effect" are also used to mean converse piezoelectric effect of the piezoelectric material. It should also be pointed out that the piezoelectric effect can be induced when heat or cooling is involved, in which case this phenomenon is termed thermoelectric or pyroelectric effect.

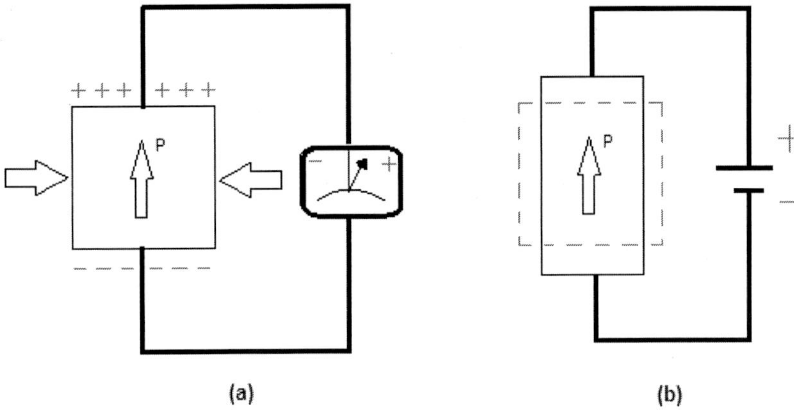

Figure 2. a) Direct piezoelectric effect; mechanical energy is converted to form electrical energy, (b) Converse piezoelectric effect; electrical energy causes deformation in the shape of piezoelectric material.

Direct piezoelectric effect is mainly used for energy harvesting. The term "Energy Harvesting" is used to describe the process of extracting energy from the environment and the extracted energy is converted and stored in the form of electrical energy.

Although energy harvesting technologies have been known for many years, increasing concern about global warming has led to intensive research for alternative energy sources including piezoelectrics. With an increasing concern about global warming, piezoelectricity has gained a significant importance and intensive research and development efforts are being made for extracting energy from the environment [Umeda et al., 1997; Sodano et al., 2004 ; Mateu & Moll, 2005].

The generated electrical charge of a piezoelectric material under a mechanical stress can be formulated in terms of dielectric displacement, D (charge per unit area, C/m^2) [Schwartz, 2002; Granstrom et al., 2007; Swallow et al., 2008]

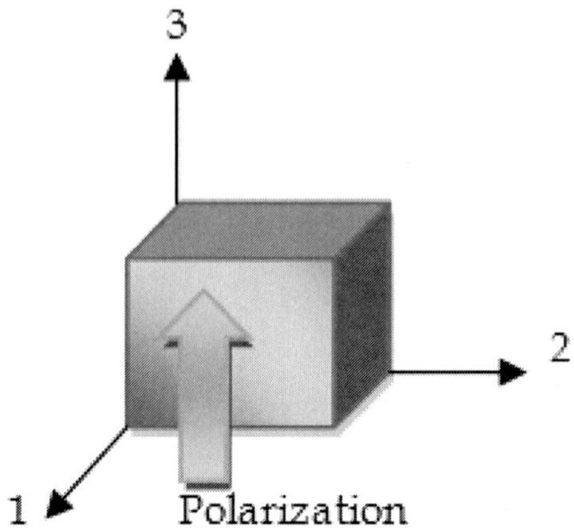

Figure 3. Defining the modes of piezoelectric material

$$D_i = d_{ij}\sigma_j \tag{1}$$

where "d_{ij}" is the piezoelectric charge coefficient (C/N) and "σ_j" is the stress (N/m^2). "i" is the direction of polarization and takes terms 1–3 and "j" is the direction of applied stress having subscripts 1–6 (Fig. 3).

In Figure 3, numbers 1, 2 and 3 present x,y and z axes of piezoelectric material, respectively. Modes 31 and 33 are two coupling modes of piezoelectric materials that when a force is applied in the perpendicular direction to the poling direction, the piezoelectric voltage coefficient acts in the mode 31. When a force is applied in the parallel direction to the poling direction, the piezoelectric voltage coefficient acts in the mode 33 which generally yields a higher coupling coefficient (k) [Roundy et al., 2005; Baker et al., 2007; Anton & Sodano 2007; Patel et al., 2010] which is the ability of piezoelectric materials to convert mechanical energy into electrical energy and vice versa, and can be defined by,

$$k = \sqrt{\frac{Y}{\varepsilon}} d \qquad (2)$$

where "ε" is dielectric constant and "*Y*" the elastic modulus of the material.

Piezoelectric voltage coefficient (*g*) relating the electric field generated by an applied mechanical stress and relationship between the piezoelectric charge coefficient and piezoelectric voltage coefficient can be expressed as [Jordan & Ounaies, 2001],

$$gij = dij / \varepsilon_0 \mathrm{Kj} \qquad (3)$$

where; " *ε0* " is the permittivity of free space (8.85x10-12 F/m) and "Kj" is the relative dielectric constant of the material.

Since the discovery of piezoelectricity in ceramics [Shirane & Suzuki, 1952; Jaffe et al., 1971] and polymers [Kawai, 1969], various studies have been carried out on structural changes [Ramos et al., 2005; Sencadas 2006], poling [Seo et al., 1985; Holstein et al., 1999; Neagu et al., 1999; Parvanova & Nadoliisky 2005] and applications, such as sensors [Tzou & Tseng 1990; Sirohi & Chopra 2000], actuators [Baz & Poh 1988; Schmidt et al., 2006], energy harvesting [Sodano & Inman 2004; Shu & Lien 2006; Granstrom et al., 2007; Ramadass & Chandrakasan 2010] and so on. PZT has been pre-eminent due to its piezoelectricity among other piezoelectric materials, with a piezoelectric coefficient (d_{33}) of 220pC/N [Hellwege, 1996] while PVDF exhibits much lower piezoelectric coefficient of $d_{33} \approx 35$pC/N [Sencadas et al., 2006 ; De-Qing, 2008; Jain et al., 2010; Patel et al., 2010]. However, flexible nature of polymers adds extra versatility for applications against ceramic based piezoelectric materials.

With the invention of new polymers exhibiting piezoelectric and better mechanical properties, the scope of application has widened. Both ceramic and polymer based piezoelectric materials have found a wide range of application in many areas. Ceramic based piezoelectric materials, in general, have a higher piezoelectric constant compared to polymer based piezoelectric materials. However, polymer based piezoelectric materials have an advantage of being flexible which makes them preferable for many applications, particularly for wearable applications. This flexibility can also result in better conversion of energy in certain applications.

3.1. Flexible piezoelectric fibres

There are many naturally occurring piezoelectric structures, such as cane sugar, tendon, silk, bones etc., but polymers are not natural piezoelectric materials. However, polymers such as polyvinylidene fluoride (PVDF), polyproplylene (PP), polyethylene terephthalate (PET), odd numbered polyamides (PA11, PA9, PA7, PA5) etc. can be made piezoelectric (Kawai, 1969; Newman et al. 1980; Dunn & Carr, 1988; Harrison & Ounaies, 2001).

Siores et al (2010) were the first to develop and patent a continuous process for making piezoelectric fibres by extruding suitable polymers. To produce piezoelectric PVDF fibres via a continuous process on the melt extruder, granular PVDF was fed to extruder screw which was heated above melting point of the polymer. To gain the piezoelectricity, molecular chain of the PVDF fibre must be re-oriented and transformed from non-piezoelectric α-phase to piezoelectric β-phase by applying a high stretching ratio (Sencadas et al., 2006), heat (Neagu et al., 1999) at the stretching region and high voltage (Holstein et al., 1999& Ramos et al., 2005), simultaneously.

3.1.1. Fibre extrusion and poling

High purity PVDF polymer granules are fed a melt extruder. The extrusion temperature is kept at 195° C which is 20° C higher than the melting point of PVDF inside the feeding screw. The temperature is slightly higher at the die, 205° C, where the fibre is extruded. The extruded fibre is then air cooled with a blower and water cooled on the initial stage rollers which help in further cooling of the extruded fibre.

Poling is a critical step for piezoelectric fibre generation. Temperature, drawing ratio and applied electric field play a crucial role in the amount of polarisation. Highest polarisation charge coefficient was given in the literature (Sessler, 1981; Wegener et al., 2002).

The drawing of fibres takes place at the rollers, which have heating coil inside to vary the temperature during stretching of fibres. The temperature of these rollers is maintained constant on PVDF fibre when it leaves the roller and an appropriate electric field applied on PVDF fibre while being drawn.

Figure 4 shows the continuous process of producing piezoelectric polymer in a customised melt extruder. This is a less expensive and less time consuming method for preparing piezoelectric polymer fibres in that all process variables are applied simultaneously. Detailed information on polymer based piezoelectric fibre production via a continuous process has been reported (Siores et al, 2010).

3.1.2. Testing of generated piezoelectric fibres

Generated polymer fibres, shown in Figure 5, are embedded in between two thin sheets of aluminium or copper which act as electrodes. The fibres are placed close to each other such that the top electrode would not contact the bottom one. The top and bottom electrodes act as positive and negative terminals for the energy generating polymer piezoelectric device.

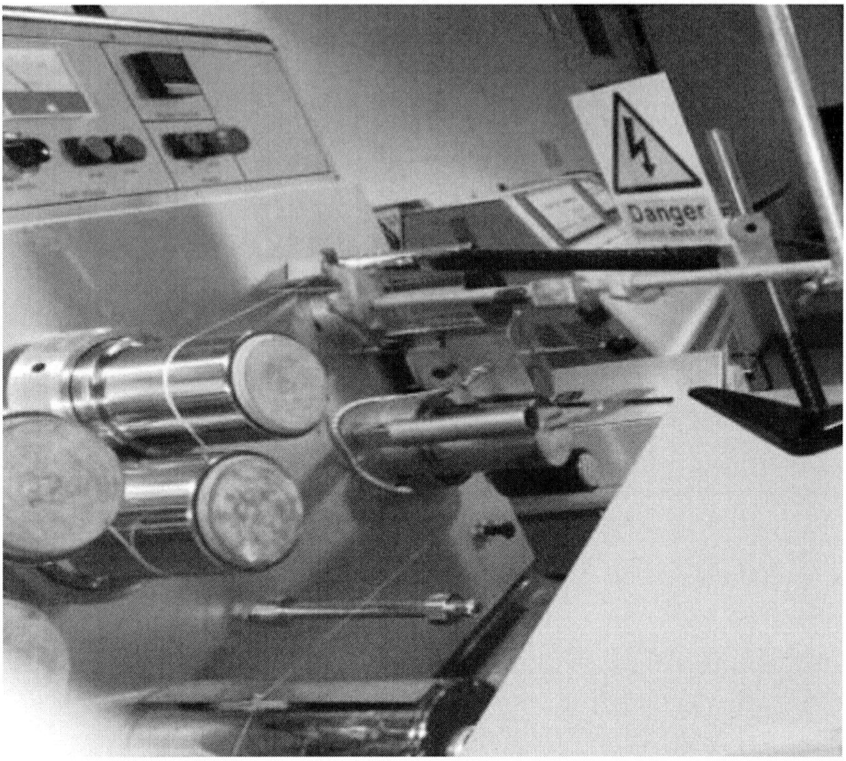

Figure 4. Piezoelectric polyvinylidene fluoride (PVDF) filament production via a continuous process using a melt extruder

Poled PVDF fibres generate about 5 V_{oc} when a moderate mechanical stimulus is applied on to the fibres. The obvious advantage of producing flexible piezoelectric fibres is to be able to produce large area active surfaces by incorporating piezoelectric fibres in wearable technologies. However, to generate enough electricity for wearable applications to power small electronic devices, produced flexible piezoelectric fibres need to be used in a fabric structure such as woven, knitted, nonwoven and 3D structures.

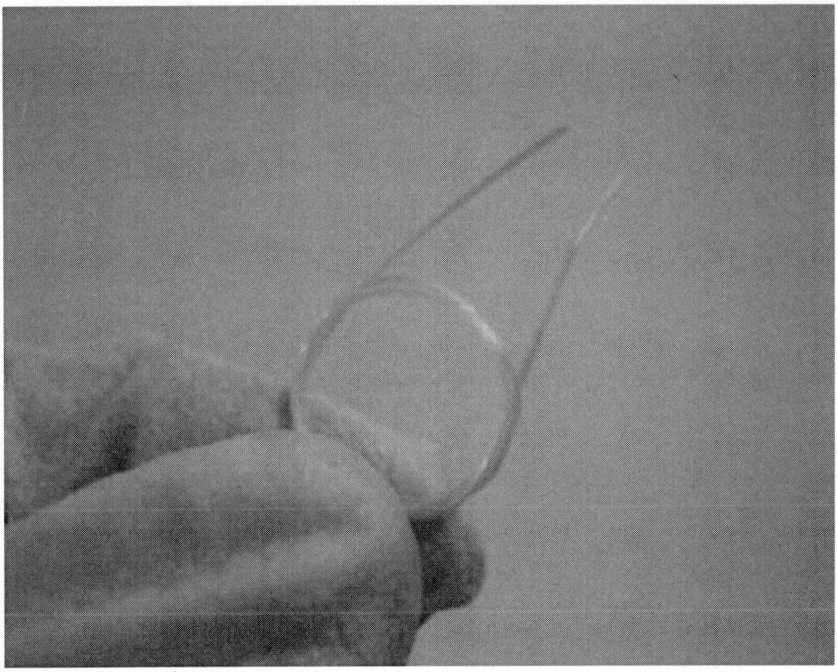

Figure 5. Polymer based flexible piezoelectric fibre generated by a continuous process using the melt extruder.

4. SMART WOVEN FABRICS

Each fabric making method has its own special attributes which help us to find most applicable fabric structure and method for a specific application. By weaving polymer based piezoelectric fibres into woven fabrics, smart piezoelectric fabrics can be produced and used for many responsive applications.

Weaving is one of the best fabric making techniques that can be used for smart fabric production. Warp threads can be located at the heddles with different orders and wefts are travelled through warps by shuttle(s). The position of heddles designates where wefts will be going over or under the warps.

While a fabric is being designed, expectations from the final fabric are taken into consideration. For smart piezoelectric fabrics, depending on expected energy generation from the final product, weaving designs can be variable. In this case, intersection of piezoelectric (the charge generator) and conductive (the charge carrier) fibres is crucial. One piezoelectric fibre can interlace more than one conductive fibre. One conductive fibre can also interlace more than one piezoelectric fibre. However, one conductive fibre can only interlace the same pole of the each piezoelectric fibre.

A number of weaving designs are studied below for smart woven fabrics. Conductive fibres and conventional (non-conductive) fibres are needed alongside piezoelectric fibres. Because piezoelectric fibres carry negative charges on one side along its length and positive charge on the other side, a conductive material is needed to carry the charge produced by movements of the piezoelectric fibres. Conductive wires would add extra rigidity to the fabric which is an undesirable outcome for most textile structures.

The best alternative to undesirable wires may be conductive fibres are produced and patented (Perera & Mauretti, 2009). It is claimed (Mauretti & Perera, 2010) that conductive filaments are flexible, non-toxic and conformable for wearable applications. Electrical conductivity of metallised synthetic (acrylic) conductive textile yarns is widely studied (Vassiliadis et al., 2004, 2009, 2010). Mechanical and electrical properties of metallised conductive yarn are controlled by blending conventional and conductive fibres in the yarn and changing the ratio of fibres in the blend. The way piezoelectric, conductive and conventional fibres are integrated into fabric structure by weaving technique, gives a good indication of the performance of resultant fabric. When more piezoelectric fibres are used in the fabric, this results in higher energy generation by movement and mechanical strain. However, to be able to carry as much charge as it is possible, the right number of conductive fibres need to interlace with piezoelectric fibres.

The possible woven fabric designs for energy generation for wearable textiles are shown in this chapter. Blue lines represent piezoelectric fibres while red lines represent conductive and grey lines show non-conductive conventional fibres. This is the simplest weaving pattern produced by plain weaving technique. However, by integrating piezoelectric and conductive fibres into this basic structure, the resultant woven fabric becomes a smart fabric which can harvest energy from the natural sources.

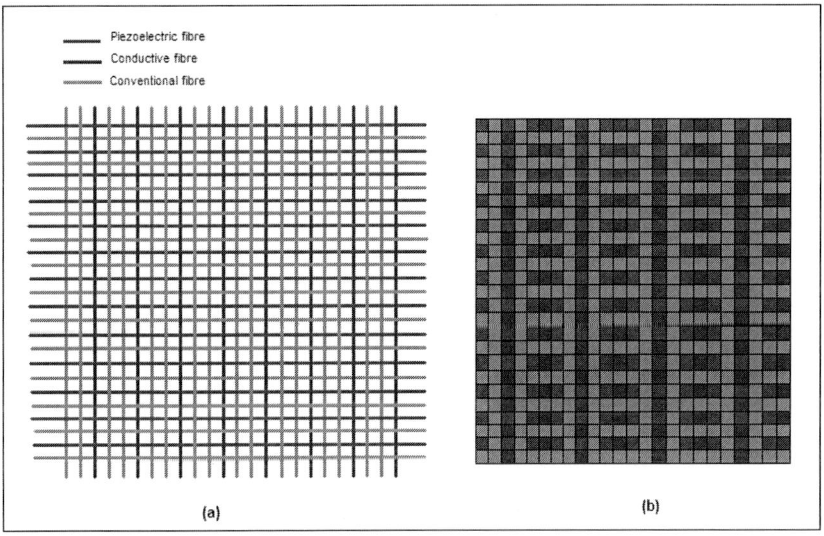

Figure 6. a) Smart woven fabric design 1 consisting of piezoelectric, conductive and non-conductive conventional fibres, (b) Face of the woven fabric consisting of piezoelectric, conductive and non-conductive fibres

Polymer based piezoelectric fibres can be used as either weft or warp into the woven structure and conductive fibres can be used as negative and positive electrodes for charge transfer so that the resultant fabric can produce energy for micro powered electronics.

The main advantage of the use of polymer based piezoelectric material in this application is its flexibility and the fact that it can easily be incorporated in the woven structures without causing any problem. It is impossible to integrate existing ceramic based piezoelectric fibres into similar structures because these fibres are rigid and brittle thus can cause major problems in the weaving process. For the first design shown in Figure 6(a), 2 heddles are needed to locate conductive and non-conductive fibres/yarns and 2 shuttles, the one with piezoelectric fibres/yarn and the other with non-conductive conventional fibres/yarn. In the warp direction, 2 conventional fibres are located between conductive fibres. Conductive fibres act as negative and positive electrodes.

If a number is given to each warp from left to right, odd numbered warps are located on the first heddle and even numbered warps are located on the second heddle. During the shuttles' travel along the loom's width, according to design, while the first conductive fibre only interlaces with negative pole of piezoelectric wefts, second conductive warp interlaces only positive pole of the piezoelectric filling fibres/yarns. Thus, any short circuit is avoided.

Figure 6(b) shows interlace of warp and weft threads and possible appearance on face of the fabric. If the used fibres counts are the same and the warps and wefts are located with an exact sequence, the resultant fabric will contain 24% piezoelectric, 16% conductive and 60% non-conductive conventional fibres/yarns.

The design shown in Figure 7(a) needs 2 heddles to locate conductive and non-conductive fibres/yarns and 2 shuttles, the one with piezoelectric fibres/yarn and the other with non-conductive conventional fibres/yarn. If a number is given to each warp from left to right, odd numbered warps are located on the first heddle and even numbered warps are located on the second heddle.

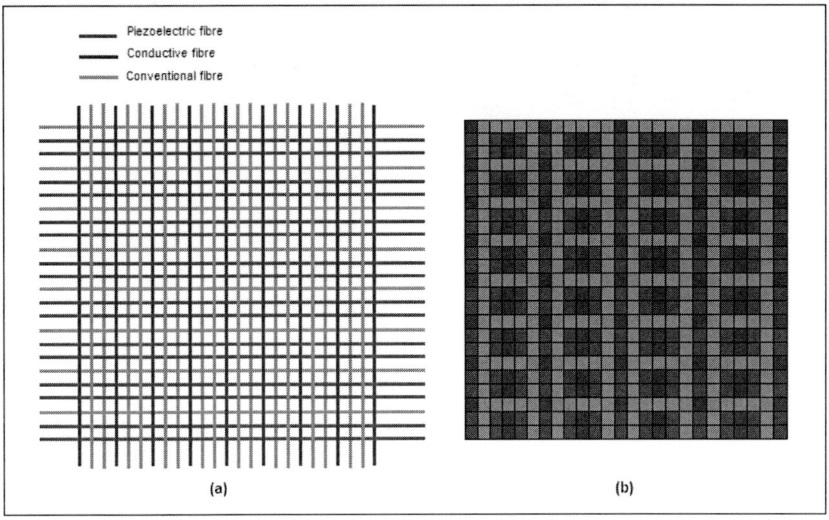

Figure 7. a) Smart woven fabric design 2 consisting of piezoelectric, conductive and non-conductive conventional fibres, (b) Face of the woven fabric consisting of piezoelectric, conductive and non-conductive fibres

During the shuttles travel along the loom's width according to the design, the first heddle is kept in place, second heddle is uplifted so that warps are kept apart and shuttle travels through easily. Shuttle carrying piezoelectric fibre travels twice and then the other shuttle which carries non-conductive conventional fibres/yarn travels once. The whole process is repeated until the desired fabric structure is created. Thus, the first conductive warp only interlaces with negative charged sides of piezoelectric wefts, second conductive warp interlaces only with the positive charged sides of the piezoelectric filling fibres/yarns.

Figure 7(b) shows interlace of warp and weft threads and possible appearance on face of the fabric. If the used fibres counts are the same and the warps and wefts are located with an exact sequence, the resulted fabric will contain 34% piezoelectric, 18% conductive and 48% non-conductive conventional fibres/yarns.

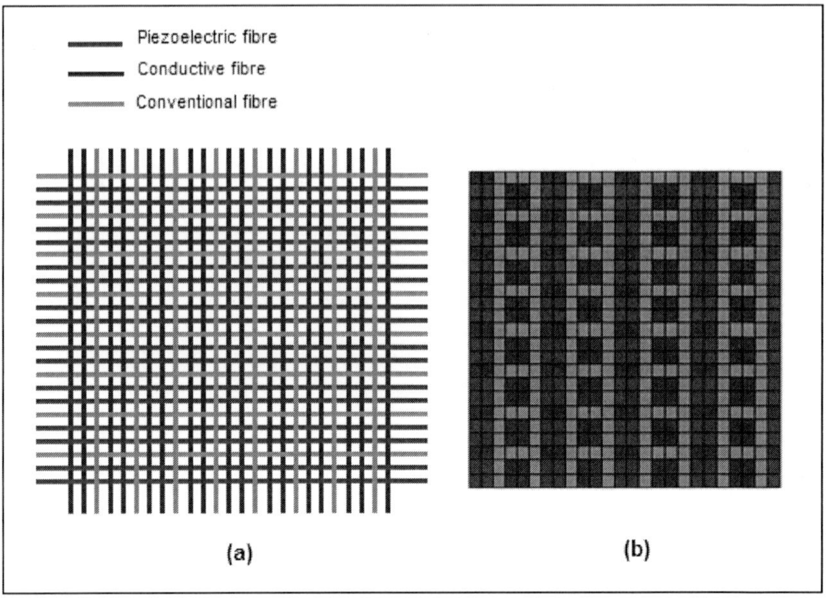

Figure 8. a) Smart woven fabric design 3 consisting of piezoelectric, conductive and non-conductive conventional fibres, (b) Face of the woven fabric consisting of piezoelectric, conductive and non-conductive fibres

The design shown in figure 8(a) needs 2 heddles to locate conductive and non-conductive fibres/yarns and 2 shuttles, the one with piezoelectric fibres/yarn and the other with non-conductive conventional fibres/yarn. If we give a number to each warp from left to right, 1st, 2nd, 7th, 8th, 13th, 14th, 19th, 20th and 25th warps are located on the first heddle and other warps are located on the second heddle.

According to design in figure 8(a), while first heddle is kept in place, second heddle is uplifted so that warps can be kept apart from the first heddle's warps and shuttle, which carries piezoelectric fibres/yarn, can easily travel through. The shuttle which carries piezoelectric fibre travels twice and then the first heddle is uplifted while the second heddle is lowered so that the other shuttle which carries non-conductive conventional fibres/yarn travels once through the warps. The same movements are carried out with the same order again and again until a fabric structure is created. Thus, all the conductive warps on the first heddle only interlace with negative pole of piezoelectric wefts and all the conductive warps on the second heddle interlace only with positive pole of the piezoelectric wefts.

Figure 8(b) shows interlace of warp and weft threads and possible appearance on face of the fabric. If the used fibres' counts are the same and the warps and wefts are located with an exact sequence, the resultant fabric will contain 34% piezoelectric, 34% conductive and 32% non-conductive conventional fibres/yarns.

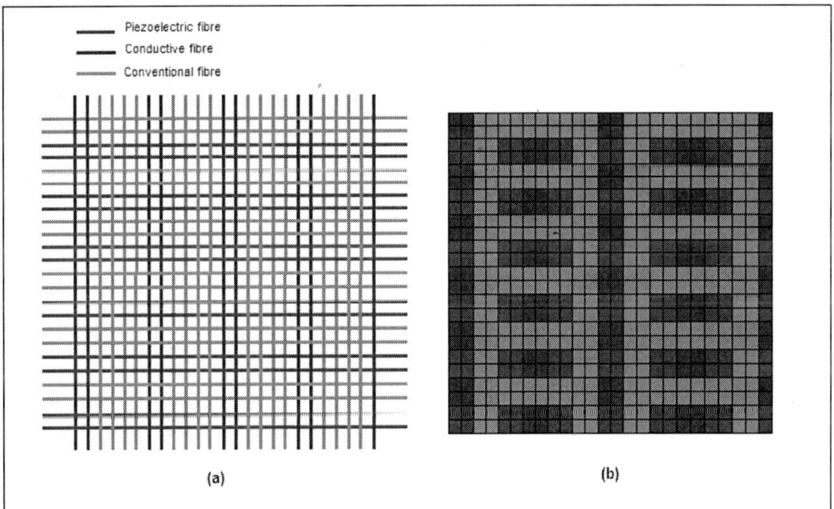

Figure 9. a) Smart woven fabric design 4 consisting of piezoelectric, conductive and non-conductive conventional fibres, (b) Face of the woven fabric consisting of piezoelectric, conductive and non-conductive fibres

The design shown in Figure 9(a) needs 2 heddles to locate conductive and non-conductive fibres/yarns and 2 shuttles, the one with piezoelectric fibres/yarn and the other with non conductive conventional fibres/yarn. If we give a number to each warp, 1st, 2nd, 5th, 6th, 9th, 10th, 13th, 14th, 17th, 18th, 21st, 22nd and 25th warps are located on the first heddle and others are located on the second heddle.

During the first shuttle's travel along the loom width, the first heddle is kept in place and the second heddle is uplifted so that warps can be kept apart from the first heddle's warps and shuttle which carries piezoelectric fibres/yarn can easily travel through. The shuttle carrying piezoelectric fibre travels twice and then the first heddle is uplifted while the second heddle is lowered so that the other shuttle which carries non-conductive conventional fibres/yarn travels twice through the warps. The same movements are carried out in the same order again and again until a fabric structure is created. Thus, all the conductive warps on the first heddle only interlace with negative pole of piezoelectric wefts and all the conductive warps on the second heddle interlace only with positive pole of the piezoelectric wefts.

Figure 9(b) shows interlace of warp and weft threads and possible appearance on face of the fabric. If the used fibres counts are the same and the warps and wefts are located with an exact sequence, the resultant fabric will contain 26% piezoelectric, 18% conductive and 56% non-conductive conventional fibres/yarns.

5. CONCLUSIONS

Polymer based piezoelectric fibres can be used as either weft or warp into the woven structure and conductive fibres can be used as negative and positive electrodes for charge transfer, therefore the resultant fabric can produce energy to power small electronic devices.

The advantage of polymer based piezoelectric fibre is their flexibility so that they can easily be used in woven structures. It is impossible to integrate existing ceramic based piezoelectric fibres into a similar structure since they are brittle.

For all four fabric designs studied in this chapter, interlace of piezoelectric and conductive fibre/yarn is significant. In a woven fabric structure, one piezoelectric fibre can interlace with more than one conductive fibre and one conductive fibre can also interlace with more than one piezoelectric fibre. However, to avoid any short circuit, one conductive fibre can only interlace with the same pole of the each piezoelectric fibre.

Since the fibres are considered having the same thickness, in the first design, piezoelectric and conductive fibres interlace 96 times in the fabric. The times of interlace of piezoelectric and conductive fibres are 153 for the second design whilst it is 289 and 117 times for the third and forth fabric designs, respectively. Therefore, the highest energy generation is expected from the third design when they all designs are subjected to the same amount of mechanical stimulus.

Smart piezoelectric woven fabrics can be used where they can be subjected to mechanical strain/stress or vibrations. Depending on the application and energy need, smart piezoelectric woven fabrics can be used to produce whole textile structure or only a part of it. For instance, tents, awnings and umbrellas can be wholly made of smart piezoelectric fabrics and produce electricity under rain as well as wind. However, waterproof finishing is needed if the fabric will be used for outdoor applications.

Energy generated by piezoelectric materials is always in the form of AC, therefore a small rectifier is needed for the conversion of the generated energy (AC) into usable energy (DC) for low power electronics.

The best weaving technique for smart piezoelectric fabric is plain weaving and its derivatives. Possible smart woven fabric designs are not limited. Depending on energy need, different fabric designs can be made with less or more interlacing. Other existing fabric making methods (other than weaving) such as embroidery can also be used to produce smart fabrics.

REFERENCES

1. S. R. Anton, H. A. Sodano, 2007 Smart Mater. Struct., 16 R1-R7.
2. J. Baker, S. Roundy, P. Wright, 2005 Proc. 3rd Int. Energy Conversion Engineering Conf., 959 970 .
3. A. Baz, S. Poh, 1988 Performance of an active control system with piezoelectric actuators, J. Sound and Vibration, 126 2 327 343 .
4. Eric. Broudy, 1979 The Book of Looms: A History of the Handloom from Ancient Times to the Present, University Press of New England, 0-87451-649-8 York, USA
5. Z. De -Qing, 2008 Chinese Physics Letters, 25 4410
6. P. E. Dunn, S. H. Carr, 1988 Piezoelectricity in nylon 5,7, IEEE Proceedings 6th International Symposium on Electrets, DOI 10.1109/ISE.1988.38559
7. D. A. Gioello, 1982 Understanding Fabrics from Fiber to Finished Cloth, Fairchild Publications, 870053779 York, USA
8. J. Granstrom, J. Feenstra, H. A. Sodano, K. Farinholt, 2007 Energy harvesting from a backpack instrumented with piezoelectric shoulder straps, Smart Materials Structures, 16 1810 1820 .
9. J. S. Harrison, Z. Ounaies, 2001 Piezoelectric Polymers, NASA/CR-2001 211422 , ICASE Report 2001-43
10. K. H. Hellwege, A. M. . Hellwege, Eds 1996 Ferroelectrics and Related Substances: Oxides, Landolt Bornstein New Series, Springer, Berlin.
11. P. Holstein, N. Leister, U. Weber, D. Geschke, H. Binder, 1999 A Combined Study of Polarization Effects in PVDF, 10th International Symposium on Electret. IEEE

12. B. Jaffe, W. R. Cook, H. Jaffe, 1971 Piezoelectric Ceramics, Academic Press, London.

13. A. Jain, J. S. Kumar, D. R. Mahapatra, H. H. Kumar, 2010 Proceedings of SPIE- The International Society for Optical Engineering 7647.

14. T. L. Jordan, Z. Ounaies, 2001 NASA ICASE Report 2001-28 2001 28 .

15. H. Kawai, 1969 The piezoelectricity of poly(vinylidene fluoride), Jpn. Journal of Applied Physics, 8 975 976

16. L. Mateu, F. Moll, 2005 J. Intell. Mater Syst. Struct., 16 835 845 .

17. G. J. Mauretti, W. R. Perera, 2010 iCon Conductive Filaments, Proceeding of Textile Institute Centenary Conference, Manchester, UK, November 3-4, 2010

18. E. R. Neagu, J. S. Hornsby, D. K. Das-Gupta, 1999 Analysis of Polarization ans Space Charge in Thermally Poled PVDF, 10th International Symposium on Electrets, IEEE, 87 90

19. B. A. Newman, P. Chen, K. D. Pae, J. I. Scheinbeim, 1980 Piezoelectricty in Nylon 11, Journal of Apllied Physics, 51 10 5161 5164 , DOI 10.1063/1.327283

20. V. D. Parvanova, M. M. Nadoliisky, 2005 Polarization Processes in PZT Ceramics, Bulg. J. Phys., 32 45 50 .

21. W. R. Perera, G. J. Mauretti, 2009 Microwires, Methods for Their Production, and Products Made Using Them, US Patent 20090260848 A1

22. Y. K. Ramadass, A. P. Chandrakasan, 2010 An Efficient Piezoelectric Energy Harvesting Interface Circuit Using a Bias-Flip Rectifier and Shared Inductor, IEEE Journal of Solid-State Circuits, 45 1 189 204 .

23. M. M. D. Ramos, H. M. G. Correia, S. Lanceros-Mendez, 2005 Atomistic modelling of processes involved in poling of PVDF, Computational Materials Science, 33 230

24. S. Roundy, P. K. Wright, J. Rabaey, 2003 Computer Communications, 26 1131 1144 .

25. V. H. Schmidt, L. Lediaev, J. Polasik, 2006 Piezoelectric Actuators Employing PVDF Coated with Flexible PEDOT-PSS Polymer Electrodes, IEEE Transactions on Dielectrics and Electrical Insilation, 13 5 1140 1148 .

26. M. Schwartz, 2002 Encyclopaedia of Smart Materials 1-2 , 780 792 .

27. V. Sencadas, V. M. Moreira, S. Lanceros-Mendez, A. S. Pouzada, R. Gregorio Jr, 2006 α-to-β Transformation on PVDF Films Obtained by Uniaxial Stretch, Materials ScienceForum 514-516, 872 876

28. V. Sencadas, R. G. Filho, S. Lanceros-Mendez, 2006 J. Non Cryst. Solids 352 2226

29. J. W. Seo, K. S. Ryoo, H. S. Lee, 1985 Characteristics of Charge Traps and Poling Behaviour of Poly(Vinylidene Fluoride), Bulletin of Korean Chemical Society, 6 4 218 221 .

30. G. M. Sessler, 1981 Piezoelectricity In Polyvinylidene fluoride, J. Acoust. Soc. Am., 70 1596 1608 .

31. G. Shirane, K. Suzuki, 1952 Crystal structure of Pb(Zr-Ti)O3, J. Phys. Soc. Jpn., 7

32. Y. C. Shu, I. C. Lien, 2006 Analysis of Power Output for Piezoelectric Energy Harvesting Systems, Smart Materials Structures, 15 1499 1512 .

33. E. Siores, R. L. Hadimani, D. Vatansever, 2010 Piezoelectric Polymer Element & Production Method & Apparatus Therefor. GB Patent Application 1015399 7

34. J. Sirohi, I. Chopra, 2000 Fundamental Understanding of Piezoelectric Strain Sensors, J. Intelligent Mater. Systems and Struc., 11 246 257 .

35. H. A. Sodano, D. J. Inman, G. Park, 2004 Shock Vib. Digest, 36 197 205 .

36. H. A. Sodano, D. J. Inman, 2004 A Review of Power Harvesting from Vibration Using Piezoelectric Materials, The Shock and Vibration Digest, 36 3 197 205 .

37. L. M. Swallow, J. K. Luo, E. Siores, I. Patel, D. Dodds, 2008 Smart Mater. Struct., 17

38. H. S. Tzou, C. I. Tseng, 1990 Distributed piezoelectric sensor/actuator design for dynamic measurement/control of distributed parameter systems: A piezoelectric finite element approach, J. Sound and Vibration, 138 17 34 .

39. M. Umeda, K. Nakamura, S. Ueha, 1997 Jpn. Journal of Applied Phys., 36 3146 3151 .

40. S. Vassiliadis, Ch. Provatidis, K. Prekas, M. Rangoussi, 2004 Electrically Conductive Spun Yarns, Proc. of the Xth International Izmir Textile and Apparel Symposium, 37 49 , Izmir, Turkey

41. S. Vassiliadis, M. Rangoussi, D. Meimaris, K. Prekas, 2009 Provaditis Ch.Electrically Conductive Spun Yarns and their Contact Behaviour, Proceedings of the International Conference on Intelligent Textiles and Mass Customisation, Casablanca, Morocco

42. S. Vassiliadis, K. Prekas, M. Rangoussi, K. Absalon, J. Maillard, 2010 The Conductive Spun Yarns as Electrical Components, Proceedings of the XIIth International Izmir Textile and Apparel Symposium, 333 338 , Izmir, Turkey

43. M. Wegener, W. Künstler, K. Richter, R. Gerhard-Multhaupt, 2002 Ferroelectric polarization in stretched piezo- and pyroelectric poly(vinylidene fluoride-hexafluoropropylene) copolymer films, J. Appl. Phys., 92 7442 7447 .

44. Wikipedia, online access from http://en.wikipedia.org/wiki/Loom#Handloom (Jan. 2011)

CHAPTER 4

Wearable Electronics and Smart Textiles: A Critical Review

Matteo Stoppa and Alessandro Chiolerio *

Center for Space Human Robotics, Istituto Italiano di Tecnologia, Corso Trento 21, 10129 Torino, Italy

ABSTRACT

Electronic Textiles (e-textiles) are fabrics that feature electronics and interconnections woven into them, presenting physical flexibility and typical size that cannot be achieved with other existing electronic manufacturing techniques. Components and interconnections are intrinsic to the fabric and thus are less visible and not susceptible of becoming tangled or snagged by surrounding objects. E-textiles can also more easily adapt to fast changes in the computational and sensing requirements of any specific application, this one representing a useful feature for power management and context awareness. The vision behind wearable computing foresees future electronic systems to be an integral part of our everyday outfits. Such electronic devices have to meet special requirements concerning wearability. Wearable systems will be characterized by their ability to automatically recognize the activity and the behavioral status of their own user as well as of the situation around her/him, and to use this information to adjust the systems' configuration and functionality. This review focuses on recent advances in the field of Smart Textiles and pays particular attention to the materials and their manufacturing process. Each technique shows advantages and disadvantages and our aim is to highlight a possible trade-off between flexibility, ergonomics, low power consumption, integration and eventually autonomy.

KEYWORDS

smart textiles; electronic textiles; wearable systems

1. INTRODUCTION

The term "Smart Textiles" refers to a broad field of studies and products that extend the functionality and usefulness of common fabrics. Smart Textiles are defined as textile products such as fibers and filaments, yarns together with woven, knitted or non-woven structures, which can interact with the environment/user. The convergence of textiles and electronics (e-textiles) can be relevant for the development of smart materials that are capable of accomplishing a wide spectrum of functions, found in rigid and non-flexible electronic products nowadays. Smart Textiles will serve as a means of increasing social welfare and they might lead to important savings on welfare budget. They integrate a high level of intelligence and can be divided into three subgroups:

- Passive smart textiles: only able to sense the environment/user, based on sensors;
- Active smart textiles: reactive sensing to stimuli from the environment, integrating an actuator function and a sensing device;
- Very smart textiles: able to sense, react and adapt their behavior to the given circumstances.

Sensors provide a nervous system to detect signals, thus in a passive smart material, the existence of sensors is essential. The actuators act upon the detected signal either autonomously or from a central control unit [1]; together with the sensors, they are the essential element for active smart materials. Fabric-based sensing has been a large field of research in the biomedical and safety communities [2]. The fabric sensors can be used for electrocardiogram (ECG) [3], electromyography (EMG) [4], and electroencephalography (EEG) [5,6] sensing; fabrics incorporating thermocouples can be used for sensing temperature [7]; luminescent elements integrated in fabrics could be used for biophotonic sensing [8]; shape-sensitive fabrics can sense movement, and can be combined with EMG sensing to derive muscle fitness [9]. Carbon electrodes integrated into fabrics can be used to detect specific environmental or biomedical features such as oxygen, salinity, moisture, or contaminants [10,11].

Active functionality could include power generation or storage [12], human interface elements [13], radio frequency (RF) functionality, or assistive technology [14]. All electronic devices require power, and this is a significant design challenge for Smart Fabrics. Power generation can be achieved through piezoelectric [15] elements that harvest energy from motion or photovoltaic elements [16]. Human interfaces to active systems can be roughly grouped into two categories: input devices and annunciation or display devices. Input devices can include capacitive patches that function as pushbuttons [17], or shape-sensitive fabrics [18] that can record motion or flexing, pressure, and stretching or compression. Annunciation and display devices may include fabric speakers [17], electroluminescent yarns [19], or yarns that are processed to contain arrays of organic light emitting diodes (OLEDs) [20]. Fabrics can also include elements that provide bio-feedback [21] or simply vibrate. Fabric-based antennas are a relatively simple application of Smart Fabrics. Simple fabric antennas are merely conductive yarns of specific lengths that can be stitched or woven into non-conducting fabrics [22].

A study about intelligent textiles is at his first stage reduced to a study on smart materials. In a second phase, it is to be considered in which way these smart materials can be processed into a textile material. These smart materials are incorporated into the textile structure by different technologies (Figure 1). Among those we may list embroidering [23], sewing, non-woven textile, knitting [24], weaving [25], making a spinning [26], braiding [27], coating/laminating [28], printing [29] and chemical treatments [30] that provide specific features such as controlled hydrophobic behavior.

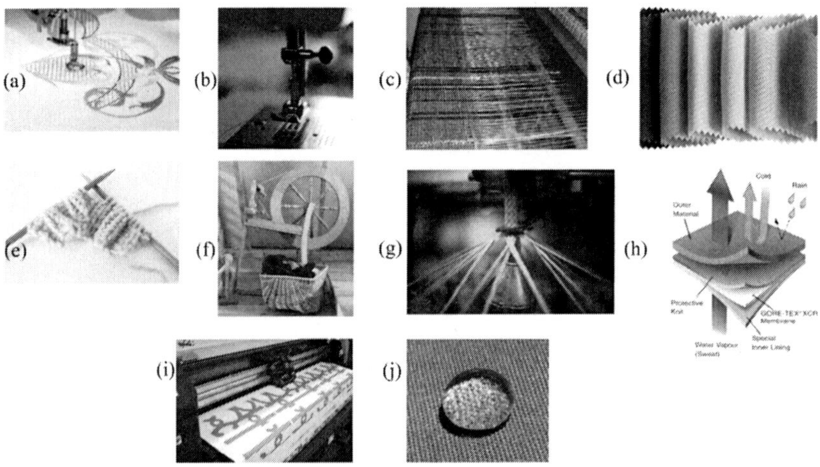

Figure 1. Different kinds of textile/fabric manufacturing and treatment. (**a**) Embroidery; (**b**) sewing; (**c**) weaving; (**d**) non-woven; (**e**) knitting; (**f**) spinning; (**g**) breading; (**h**) coating/laminating; (**i**) printing and (**j**) chemical treatment.

Innumerable combinations of these source materials result into a whole range of textiles but sometimes the commercial output is represented by garments that contain conventional cables, miniaturized electronic components and special connectors. As humans prefer to wear comfortable textiles rather than hard, rigid boxes, first efforts have been made to use the textiles themselves for electronic functions [31].

Smart Textiles present a challenge in several fields such as the medical, sport, and artistic communities, the military and aerospace. The early European Commission's 6th and 7th framework programs provided significant research and development funding for personal health monitoring through smart wearable systems and for projects targeting the integration of sensors, energy sources, processing, and communication inside the clothing. The list below shows the projects funded by the European Commission's 6th and 7th Framework programs that have focused on smart fabrics and interactive textiles (Table 1) [32,33].

Particular attention in this review is devoted to describing the materials and methodologies to develop smart textiles. Each scientific approach will be followed by a review of the related work carried out by companies, universities or research institutes.

2. FABRICATION TECHNIQUES

Over the past decade, many techniques and materials have been used in order to realize smart textiles. In the following section, together with metodologies, the relative projects are also presented.

2.1. Conductive Fibers

Initially, conductive threads were mainly used in technical areas: clean room garments, military apparel, medical application and electronics manufacturing [34]. Textile structures that exhibit conductivity or serve an electronic or computational function are called electro-textiles [35]. They can have a variety of functions, like antistatic applications [36], electromagnetic interference shielding (EMI) [37], electronic applications, infrared absorption or protective clothing in explosive areas [38].

Table 1. Smart Textiles project—EU FP6 and FP7.

Project Title	Description
WEALTHY Sept 2002–Feb 2005 http://www.wealthy-ist.com/	Pioneering research to develop and test comfortable smart fabrics for biological monitoring–ECG and respiration.
MyHeart Jan 2004–Oct 2007 http://www.hitech-projects.com/euprojects/myheart/	Development of Intelligent Biomedical Garments for monitoring, diagnosing and treatment of medical conditions
BIOTEX Oct 2005–Feb 2008 http://www.biotex-eu.com/	Sought to develop biochemical-sensing techniques that could integrate into textiles. Project aimed to develop textile sensing patches to target bodily fluid sensing.
PROETEX Feb 2006–Jan 2010 http://www.proetex.org/	Developed smart wearable garments for emergency disaster intervention personnel to improve safety, coordination, and efficiency.
STELLA Feb 2006–Jan 2010 http://www.stella-project.de/	Sought to develop flexible and stretchable textile substrates with electrical interconnects.
OFSETH Mar 2003–Jun 2009 http://www.ofseth.org/	Focused on how silica and polymer optical fibers can be used for sensing vital parameters while being compatible with a textile manufacturing process.
CONTEXT Jan 2006–Jun 2008 http://www.context-project.org/	Focused on development of contactless sensors in textiles for monitoring ECG and EMG.
WearIT@Work Jun 2004–Nov 2008 http://www.wearitatwork.com/	Aimed to prove the applicability of computer systems integrated to clothes, creating wearable interfaces for various industrial environments.
MICROFLEX May 2008–May 2012 http://microflex.ecs.soton.ac.uk/	Development of flexible materials in the form of high added value smart fabrics/textiles which are able to sense stimuli and react or adapt to them in a predetermined way.
DEPHOTEX Nov 2008–Oct 2011 http://www.dephotex.com/	Research and development of flexible photovoltaic textiles based on novel fibers
PLACE-it Feb 2010–Jun 2013 http://www.place-it-project.eu/	Development of a technology platform for lightweight, thin and conformable opto-electronic systems interconnect technology

The conventional process to produce metal fibers is wire drawing, a mechanical production process. This process is characterized by its various drawing steps, called coarse, medium, fine and carding train (Figure 2).

The drawing die, used to draw the fiber, consists of a steel mount with a core out of ceramics, carbide or diamond. The initial diameter of the metal wire varies depending on the material. For copper, for instance, it is usually is 8 mm, while for iron it is 5 mm. After drawing, the wire is annealed at temperatures ranging between 600 and 900 °C. Subsequently, they are quenched. The fine metal wire is then wrapped onto a revolving wire drawing cylinder [39].

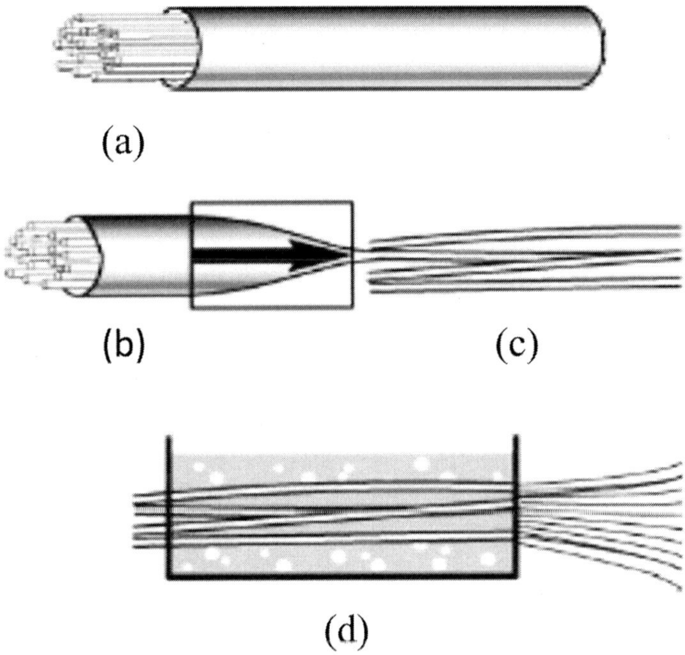

Figure 2. (**a**) Metal coated wire combined in iron tube; (**b**) Several diameter reductions of tube; (**c**) Bundling of tubes; (**d**) Leaching, realizing fibers.

Related Works

The Swiss company Elektrisola Feindraht AG (Escholzmatt, Switzerland) produces metal monofilaments that can be blended with all sorts of fibers or that can be directly used in weaving and knitting. Importantly, according to the material used, there are different electrical properties (see Table 2). The products range from copper (Cu) and silver-plated copper (Cu/Ag) filaments, brass (Ms) and silver-plated brass (Ms/Ag) filaments, aluminum (Al) filaments to copper-clad aluminum (CCA) filaments [40].

Table 2. Electrical properties of metal monofilaments fibers.

Metal	Electrical Properties				
	Conductivity [S·m/mm²]	Resistivity [Ω·mm²/m]	Thermal Coefficient of Resistance [10^{-6} K^{-1}]		
			Min	Typ	Max
Cu	58.5	0.0171	3900	3930	4000
Cu/Ag	58.5	0.0171	3900	4100	4300
Ag 99%	62.5	0.0160	3800	3950	4100
Ms * 70	16.0	0.0625	1400	1500	1600
Ms/Ag	16.0	0.0625	1400	1500	1600
AgCu	57.5	0.0174	3800	3950	4100
Bronze	7.5	0.1333	600	650	700
Steel 304	1.4	0.7300		1020	
Steel 316L	1.3	0.7500		1020	

* German Milbe denomination, where "Ms" is accompanied by a number stating the composition in %Cu with respect to a Zn complement to 100%.

The company Swiss-Shield® (Flums, Switzerland) specializes in producing metal monofilaments which are incorporated into base yarns like cotton, polyester, polyamides and aramides. The metal monofilaments are made out of copper, brass, bronze, silver, gold, aluminum, for instance. The following Figure 3 shows a typical conductive yarn with base fibers and a metal monofilament twisted around them [41].

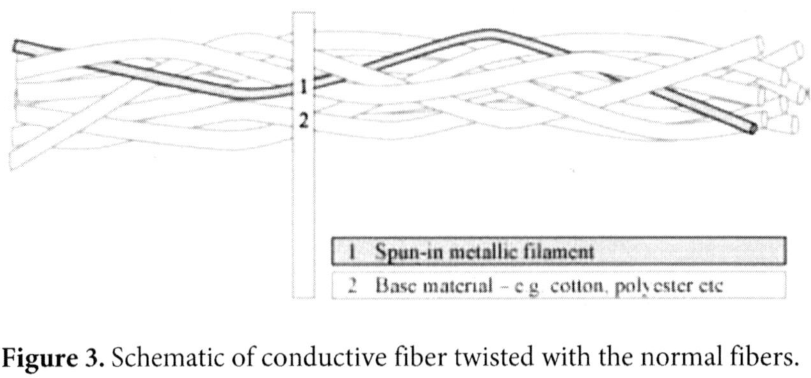

Figure 3. Schematic of conductive fiber twisted with the normal fibers.

2.2. Treated Conductive Fibers

Instead of attaching electronics to textile substrates, the yarns of the textile can be functionalized with electronics. Electrically conductive fibers can also be produced by coating the fibers with metals, galvanic substances or metallic salts. Coatings can be applied to the surface of fibers, yarns, or even fabrics to create electrically conductive textiles. Common textile coating processes include electroless plating, evaporative deposition, sputtering, coating the textile with a conductive polymer [42].

In [43] a method to fabricate fibers with different material layers and material structuring is presented. The fabrication process is based on the conventional preform-based fiber-processing, easily yielding kilometers of functional fiber during the process.

Another relevant work is to use the crossing yarns in a textile to fabricate a transistor [44,45]. A schematic of a yarn-based transistor is shown in Figure 4. The resulting transistor shows an on-off current ratio of more than 1000, operated with a gate voltage of 1.5 V.

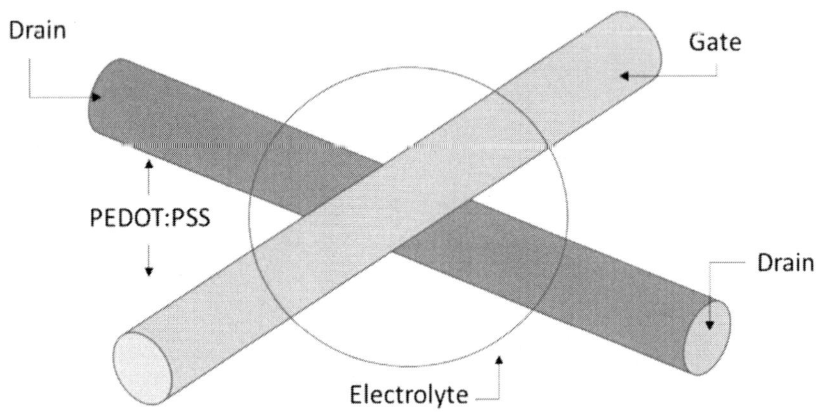

Figure 4. Yarn-based transistor.

Figure 3 represents two yarns coated with PEDOT:PSS, one serving as the gate contact for the transistor while the second serves as drain and source contact. At the crossing of the two yarns, an electrolyte is placed. A redox process at the interface between electrolyte and PEDOT:PSS turns the transistor on and off [44].

Related Works

The Textile Research Institute of Thuringia-Vogtland (TITV, Greiz, Germany) has succeeded in producing conductive threads by coating a conventional yarn with metal layers, called ELITEX®. They used Shieldex Nylon 66 threads that are coated with a thin silver layer as base material. With a specific conductivity of about 1.2×10^3 $S{\cdot}cm^{-1}$, the threads have a specific resistance of about 8.34 $O{\cdot}mm^2/m$. Hence, the resistivity is too high to conduct current [46].

2.3. Conductive Fabrics

There are different ways to produce electrically conductive fabrics. One method is to integrate conductive yarns in a textile structure, e.g., by weaving. However, the integration of conductive yarns in a structure is a complex and seldom a uniform process as it needs to be ensured that the electrically conductive fabric is comfortable to wear or soft in touch rather than hard and rigid. Conductivity can be established with different thread types (Figure 5):

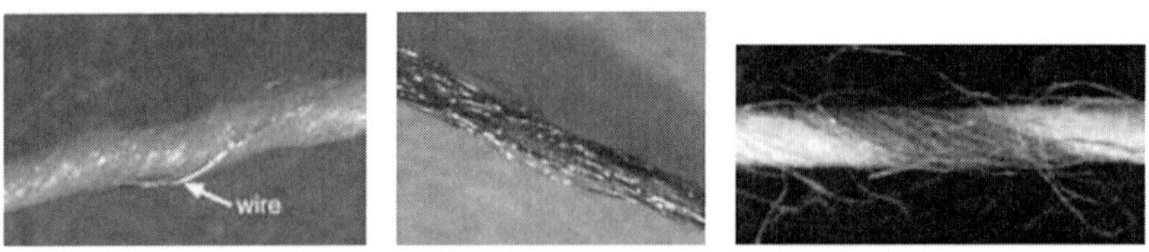

(a) (b) (c)

Figure 5. (**a**) Twisted metal wire: The metal wire is twisted around the polymer yarn; (**b**) Metal coating: The polymer yarn is physically/chemically coated with a thin metal layer; (**c**) Metal fibers: The conductive yarn consists of metal multifilaments [47].

However, woven fabric structures can provide a complex network that can be used as elaborated electrical circuits with numerous electrically conducting and non-conducting constituents, and be structured to have multiple layers and spaces to accommodate electronic devices.

Researchers at the Electronics Department and the Wearable Computing Laboratory at the ETH in Zürich produced a plain woven textile structure consisting of polyester yarns that are twisted with one copper thread. Initially, they started with a standard design (Figure 6a), then the researchers design a hybrid fabric called PETEX (Figure 6b) [47]. It consists of woven polyester monofilament yarn (PET) with diameter of 42 μm and copper alloy wires with diameter 50 ± 8 μm (AWG 461). Each copper wire itself is coated with a polyurethane varnish as electrical insulation. The copper wire grid in the textile features a spacing of 570 μm (mesh count in warp and in weft is 17.5 cm^{-1}).

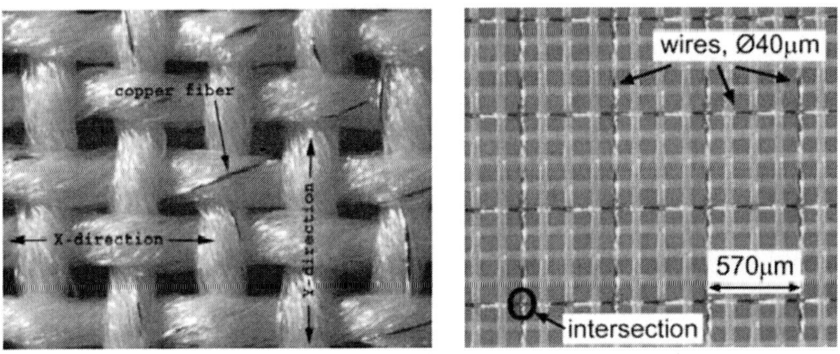

(a) (b)

Figure 6. (**a**) Standard design of copper yarn twisted with polyester fibers; (**b**) PETEX.

With the PETEX the ETH researchers introduced a new approach to Smart Textiles and in particular a new manufacturing method. The aim was the possibility to realize a custom textile circuit (Figure 7). The wiring structure among circuit components is established by connecting the fabric embedded copper wires. Cuts must be placed at specific locations in the wiring in order to avoid short-circuits between copper wires. In particular, the procedure is as follows [47]:

- coating removal on copper wires at defined intersections with laser ablation;
- cutting the wires avoiding the signal leakage with laser;
- creating the interconnection with a drop of conductive adhesive;
- adding mechanical and electrical protection with an epoxy resin deposition.

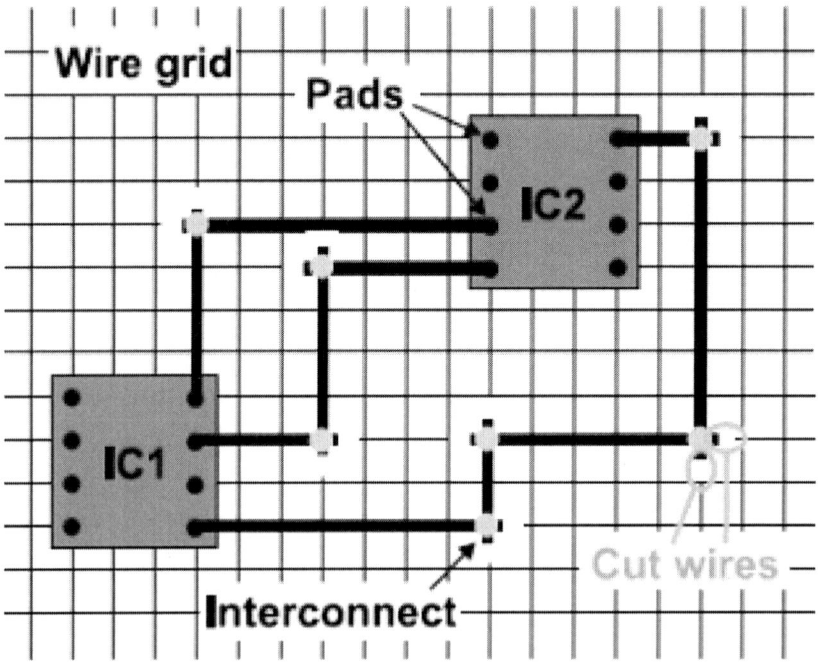

Figure 7. Approach to integrate circuits in a fabric with wire grid.

Related Works

The British company Baltex (Ilkeston, UK) uses the knitting technology to incorporate metal wires in textile structures. Their fabrics, which they commercialize under the name Feratec®, can be used mainly for two purposes, namely heatable textiles and electro-magnetic shielding materials [48].

The American company Thremshield LLC (Niagara Falls, NY, USA) produces metallized woven nylon fabrics in different shapes and profiles. The metals they use are silver, copper or a combination of copper and nickel [49].

The Danish company Chr. Dalsgaard Project Development (Aarhus, Denmark) works with the development of weaving electronics into fabrics, electronic conductors in clothing, operating panels in textiles (soft keyboards, displays, etc.) and micro-sensors. The conductive yarn they use is a copper thread, plated with a silver layer and coated with polyester [50].

Another possibility to achieve a conductive fabric is to attach a conductive structure to a ground structure by using the embroidery technique. In 2000, the Massachusetts Institute of Technology Media Laboratory researches were the first to propose a way of stitching patterns that can define circuit traces, component connection pads, or sensing surfaces designed with traditional CAD tools for circuit layout (Figure 8) [31].

(a) (b)

Figure 8. (**a**) Musical Jacket comprising a fabric keypad on one side, a MIDI synthesizer on the other side, speakers behind speaker grills in the pockets and fabric buses visible inside the jacket; (**b**) The fabric keypad with the circuit board placed behind it.

2.4. Conductive Inks

Interactive electronic textiles can also be produced by using conductive inks. First of all conductive inks must contain an appropriate highly conductive metal precursor such as Ag, Cu, and Au NPs and a carrier vehicle. Most of them are water based: water is the main ink component and to limit contaminants, it must be as pure as possible. These specialized inks can be printed onto various materials, among them textiles, to create electrically active patterns. Screen printing also makes integration with planar electronics simpler than with conductive yarn systems.

There are several technologies that can print conductive material on different substrate. Sheet-based inkjet and screen printing are best for low-volume, high-precision work.

Inkjets are flexible and versatile, and can be set up with relatively low effort [51]. Inkjets offer lower throughput of around 100 m^2/h and lower resolution (ca. 50 μm). It is well suited for low-viscosity, soluble materials like organic semiconductors. With high-viscosity materials, like organic dielectrics, and dispersed particles, like inorganic metal inks, difficulties due to nozzle clogging occur. Because ink is deposited via droplets, thickness and dispersion homogeneity is reduced. Simultaneously using many nozzles and pre-structuring the substrate allows improvements in productivity and resolution, respectively [52].

For inkjet printing, the inks should respect the following requirements [53]:

- high electrical conductivity;
- resistance to oxidation;
- dry out without clogging the nozzle during printing;
- good adhesion to the substrate;
- lower particle aggregation;
- suitable viscosity and surface tension.

Inks may also contain additives which are used to tune ink properties or to add specific properties thus increasing its performance [54].

After inkjet printing of a metal NP-based ink, in order to form a conductive printed pattern, particles must be sintered to create continuous connectivity between them and obtain electrical percolation. Sintering is the process of welding particles together at temperatures below the corresponding bulk metal melting point, involving surface diffusion phenomena rather than phase change between the solid and the liquid [55]. For instance, with inks based on gold NPs (1.5 nm diameter), the melting temperature was experimentally found to be as low as 380 °C, while for inks based on silver NPs (15 to 20 nm in diameter), a complete sintering can be obtained down to 180 °C [54,56].

Screen printing is appropriate for fabricating electrics and electronics due to its ability to produce patterned, thick layers from paste-like materials. The screen printing procedure, a stencil process, comprises the printing of a viscous paste through a patterned fabric screen and is usually followed by a drying process. The method can be applied to flat or cylindrical substrates. Depending on the substrate materials and the requirements for the printed structures, a high temperature densification can also be necessary (organic substrate $T < 150$ °C. Glass, ceramic and metal substrate $T > 500$ °C) [57].

This method can produce conducting lines from inorganic materials (e.g., for circuit boards and antennas), but also insulating and passivating layers. Generally, the throughput is about 50 m^2/h with a resolution lower than 100 μm. By optimizing of the process condition and material the resolution may decrease to 30 microns line/space on thin flexible substrates [58].

This versatile and comparatively simple method is used mainly for conductive and dielectric layers [59], but also organic semiconductors can be printed, e.g., for OPVCs and OFETs [60].

Recently the researchers of University of Southampton [61] have developed an innovative screen printed network of electrodes and associated conductive tracks on textiles for medical applications (Figure 9). A polyurethane paste is screen printed on to a woven textile to create a smooth, high surface energy interface layer and a silver paste is subsequently printed on top of this interface layer to provide a conductive track. Silver pastes have been printed on to non-woven textiles to create wearable health monitoring devices.

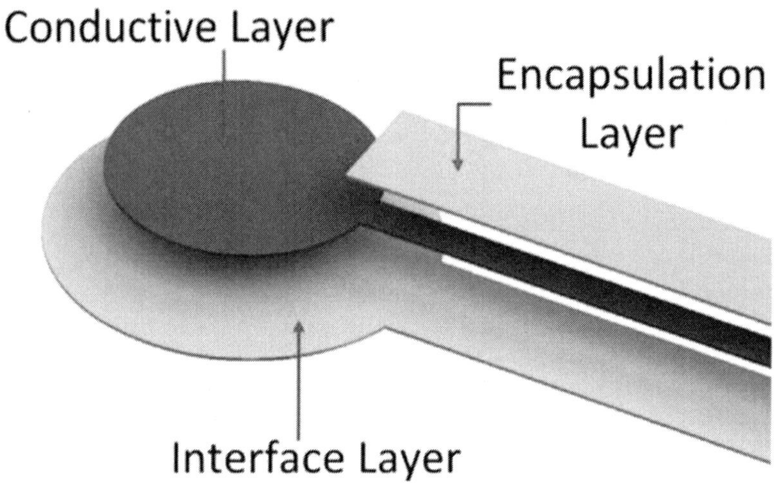

Figure 9. Screen printing fabrication for conductive tracks.

The researchers developed different bio-potential sensing systems with dry electrodes and conductive ink for signal traces in order to demonstrate that this technology could be used for biomedical application. In particular the application tests were: ECG, facial EMG (Figure 10) and forearm EMG.

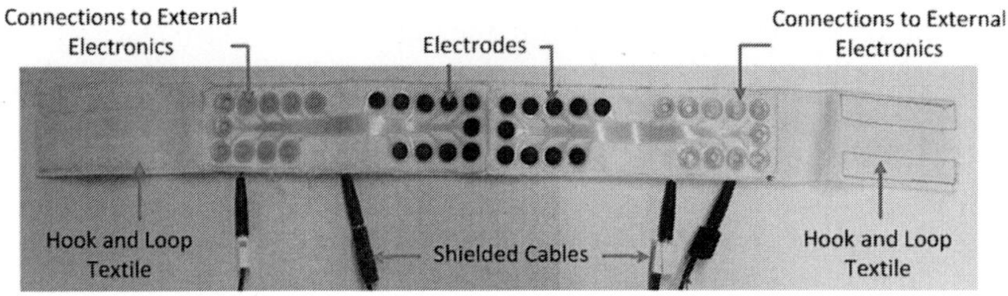

Figure 10. Textile headband for facial EMG.

Related Works

The National Textile Center of the North Carolina State University (Raleigh, NC, USA) is currently working on a project dealing with 'Printing Electric Circuits on Non-Woven Fabrics' used to produce a prototype for a physiological monitoring garment that measures ECG, heart-rate, respiration and temperature. In the scope of the project they work together with conductive ink manufacturers. For their experimental investigations and to succeed in producing samples of antennas printed on non-woven textile structures, they used Evolon® by Freudenberg KG (Weinheim, Germany), Tyvek® by DuPont™ (Wilmington, DE, USA) FiberWeb Resolution™ Print Media by BBA FiberWeb™ (Old Hickory, TN, USA) as well as conductive inks made by Precisia LLC (Clark, NJ, USA) and Creative Materials Inc. (Ayer, MA, USA) [62]. Table 3 shows the ink's sheet resistivity.

Table 3. Electrical properties of inks.

Inks	Sheet Resistivity (Ω/sq/mil)
CMI 112-15	0.010
DuPont 5025	0.010–0.012
DuPont 5096	<0.010

A group of researchers of Istituto Italiano di Tecnologia—Center for Space Human Robotics, Politecnico di Torino—Applied Science and Technology Department, in collaboration with a spin-off company, Politronica Inkjet Printing S.r.l. (Torino, Italy), developed EMG sensor matrices by inkjet printing a silver nanoparticle-based ink on a polyimide flexible patch. Results indicate excellent behavior with respect to traditional bulk silver systems and a base conductivity of 35% with respect to bulk silver [63].

2.5. Conductive Materials as Sensors

Conductive textiles that change their electrical properties as a result of the environmental impact can be used as sensors. Typical examples are textiles that react to deformations such as pressure sensors, stretch sensors and breathing sensors. On the other hand, with smart textiles we have the further possibility to make bio-potential sensors.

2.5.1. Stretch Sensors

Stretch sensors are predominantly used for sensing and monitoring body parameters, as the textile is in contact with the skin over a large body area. This means that monitoring can take place at several locations on the body. For instance, these sensors can be used for determining: heart rate, respiration, movement and pressure blood [64]. A specific structure of textile sensors is that integrating fibers featuring piezo-resistive properties, enabling their use as strain or deformation sensor.

A first approach to integrate electronics into textile structures was certainly achieved by gloves wired to the computer that allows it to take input from user's hand gestures. Sensors in the glove detect (Figure 11) the wearer's hand movements. Four wires were used for each finger or tube to build up a circuit. The voltages coming out are varying depending on the finger position [65].

Figure 11. DataGlove™ VRLOGIC with flex sensors.

Related Works

A Flemish consortium of universities and companies, among them the textile department of Ghent University, developed a prototype suit called Intellitex. It is a biomedical suit meant for the long term monitoring of heart rate and respiration of children at the hospital. To measure the ECG, a three-electrode configuration is used. Two measurement electrodes are placed on a horizontal line on the thorax, a third one, acting as a reference (Right Drive Leg, RDL), is placed on the lower part of the abdomen [66].

Philips Research Laboratory (Redhill, UK), developed a stretch sensor integrated into a garment. The stretch sensor, which is produced out of conductive and elastic yarns knitted together, is based on the fact that the electrical resistance changes when stretching the sensing material. Thus, it can be used to control the volume of music or changing the track [67].

2.5.2. Pressure Sensors

Pressure sensors are commonly used either as switches and interfaces with electronic devices or also to monitor vital signs of the user.

Several technologies [68,69], have been developed to manufacture plane pressure sensors. The operating principle is that of changes in piezoelectric resonance frequency with the applied pressure or capacitance variations caused by an elastic foam overlaid with a matrix of conductive threads. For capacitive sensors, a change in parasitic capacitance and resistance can be compensated by the electronics, therefore the wiring has a marginal influence on the sensed signal [9].

The Wearable Computing Lab of ETH Zurich has developed a matrix with several capacitive pressure sensors for integration into a piece of clothing (Figure 12). With this method they are able to measure pressure on a human body and detect muscle activity on the upper arm. Applying this matrix on different body areas, it can provide more details for motion tracking or for the detection of physical state of the muscles [9].

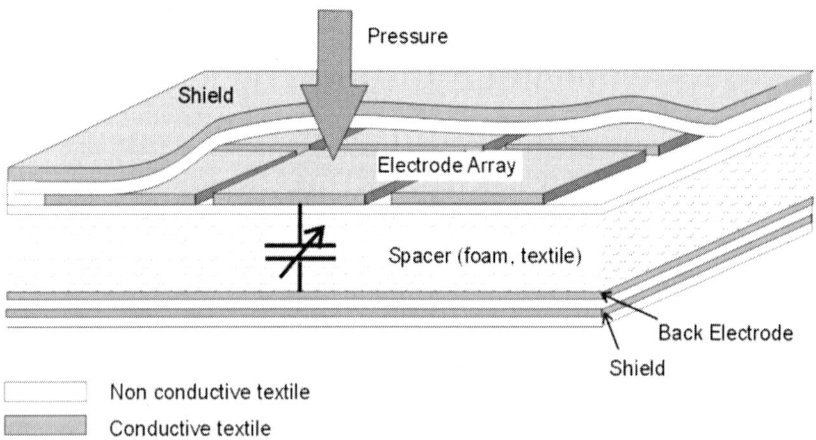

Figure 12. Scheme of sensor with an array of textile capacitors.

Related Works

The British company Eleksen Limited, formerly Electrotextiles (Tunstall, UK), commercializes a soft and flexible textile based sensory fabric under the tradename ElekTex® Smart Fabric Interfaces. It is a combination of conductive fibers and nylon. This combination results into a durable, reasonably priced, washable and even wearable 3D structure [70].

The American-based company Logitech Inc. (Le Lieu, Switzerland) manufactures a soft-touch KeyCaseTM keyboard that can wrap around a personal digital assistant (PDA) for storage and protection. The keyboard is lightweight and made out of textile [71].

The U.S. company Pressure Profile Systems, Inc. (Los Angeles, CA, USA) designs, develops and manufactures high performance multi-element pressure and tactile sensing systems, called Tactarray and ConTacts (Figure 13a) [72].

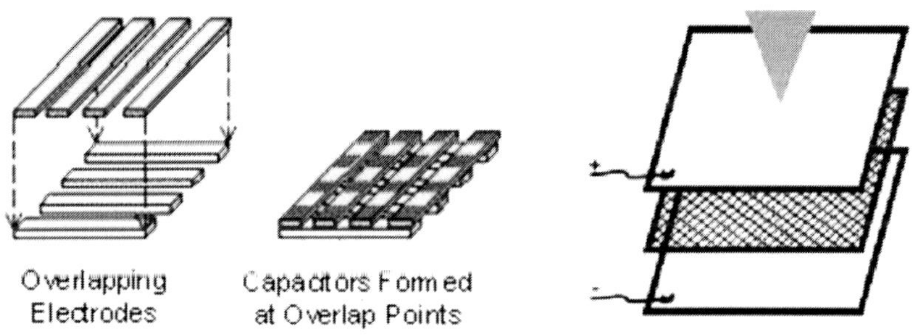

(a) (b)

Figure 13. (**a**) Scheme of capacitor sensor developed by the U.S. Company Pressure Profile Systems, Inc. and (**b**) Design for Life Centre at Brunel University in Surrey.

The team of the Design for Life Centre at Brunel University in Surrey (UK) has developed a fabric (the Sensory Fabric) that can be used by handicapped children to make themselves understood (Figure 13b). The Sensory Fabric consists of two layers of electrically conductive textile, divided by a layer of non-conductive mesh. When the textile is pressurized, one conductive layer comes into contact with the other, as a result of which an electric stream can flow [73].

The team of the Center for Micro-Bio Robotics of Istituto Italiano di Tecnologia (Pisa, Italy) produced a composite capacitive three-axial sensor based entirely on commercial conductive fabrics, demonstrating its high compliance and stability under manipulation [74].

2.5.3. Electrochemical Sensors

Recent insights into novel fabrication methodologies and electrochemical techniques have resulted in the demonstration of chemical sensors able to augment conventional physical measurements (i.e., heart rate, EEG, ECG, etc.) [75]. Recent insights have resulted in the development of a new generation of textile-based chemical sensors that are able to improve conventional physical sensors with more information. Flexible and textile-based screen printed electrochemical sensors may be candidates for non-invasive monitoring, but these devices cannot easily be attached to the body and in particular to the skin.

Related Works

The researchers of National Centre for Sensor Research, Dublin City University (Ireland), present examples of wearable chemical sensors that monitor the person and also their environment. In particular the chemical sensor was able to measure and analyze sweat in real-time on the body. They have developed a microchip version of the platform to measure changes in the pH of sweat (Figures 13a and 14a). The color change of the pH sensitive fabric was detected by placing a surface mount (SMT) LED and photodiode module on either side of the chip, aligned with the pH sensitive fabric. The final device (180 μm thick) is flexible and can adapt to the body [76].

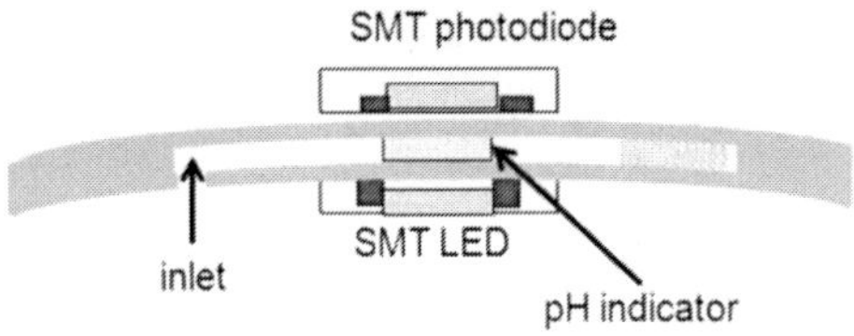

(a)

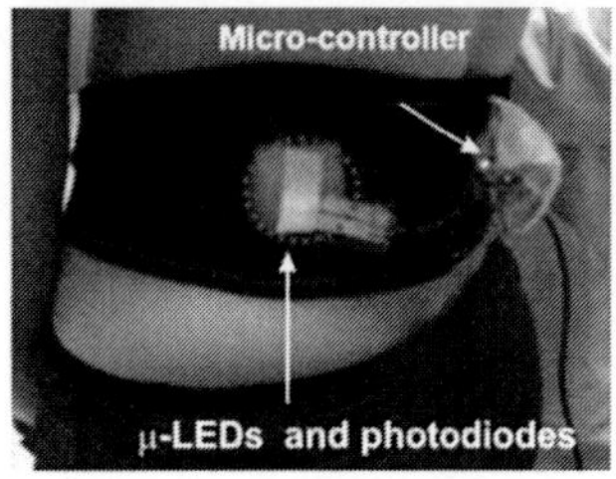

(b)

Figure 14. (**a**) Scheme of electrochemical sensor for pH analysis and (**b**) the system application.

2.5.4. Textile Energy Harvesting and Portable Power Supply System

Power supply technologies provide the electrical power for activating the components integrated in the electronic textile; this is still a critical issue in the field of wearable electronics. Although considerable progresses have been seen for wearable electronics, lithium rechargeable batteries, the power sources of the devices, have not kept pace with this progress due to their tenuous mechanical stability, causing them to remain as the limiting elements in the entire technology [77]. For these reasons, the aim is to develop wearable systems capable of accumulating energy dissipated by the body.

The supply of energy by the user's body during everyday actions through leg motions and body heat is also exploited by other research teams. For example, Infineon is currently trying, to recover energy by body movements to feed Mp3 players integrated in a jacket using piezoelectric materials [78]. In the UK the University of Bolton has developed a novel technology that integrates piezoelectric polymer substrate and photovoltaic coating system to create a film or fibre structure that is capable of harvesting energy from nature, including sun, rain, wind, wave and tide [79].

In a project funded by the Engineering and Physical Sciences Research Council (EPSRC), Researchers at the University of Southampton's School of Electronics and Computer Science (ECS) are developing an energy harvesting film in textiles using a rapid printing processes and active printed inks [80,81].

Georgia Tech researchers led by materials-science professor Zhong Lin Wang, have made a flexible fiber coated with zinc oxide nanowires that can convert mechanical energy into electricity. The researchers say the fibers should be able to harvest any kind of vibration or motion for electric current. Gold-plated zinc oxide nanowires, each about 3.5 micrometers tall, are grown on a flexible polymer fiber and these nanowires brush against untreated nanowires, which flex and generate current. Yarn spun from the fibers could lead to fabrics that convert body movements into electric current [82].

Another technology is represented by the solar clothing in which the solar energy is harvested through new generation of flexible solar cells [83,84]. Integration of flexible solar cells into clothing can provide power for portable electronic devices. Photovoltaic is the most advanced way of providing electricity far from any mains supply, although it suffers from the limits of ambient light intensity. Nevertheless the energy demand of portable devices is now low enough that clothing-integrated solar cells are able to power most mobile electronics [85].

Related Works

The ILLUM jacket (Figure 15) is based on technologies including printed electroluminescent ink and printed photovoltaic technology. The functional parts are placed outside the jacket and into several ergonomic panels, while at the front and the photovoltaic source at the shoulders and top of the back [86].

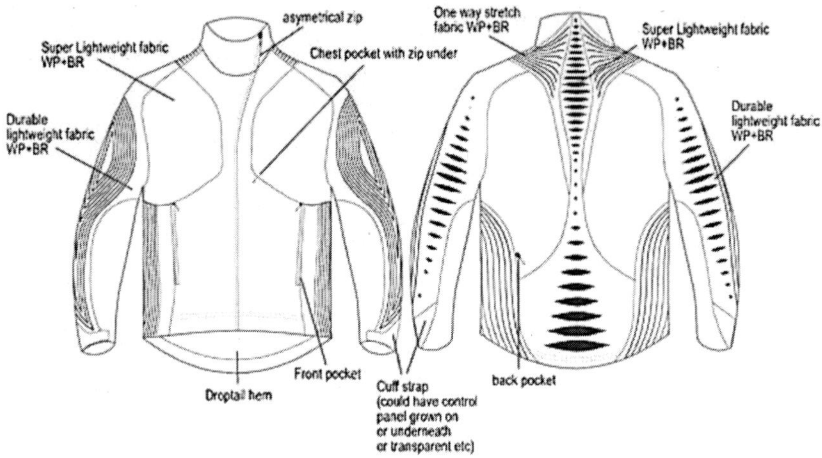

Figure 15. ILLUM jacket layout.

Thermotron of UNITIKA (Osaka, Japan) is a particular fabric able to converts sun light into thermal energy while storing heat without wasting it. Inside the Thermotron there are microparticles of zirconium carbide which allow the fabric to absorb and filter sunlight. The inner layer of the fabric withholds the heat generated and prevents it from becoming lost, thus providing a salutary effect on the human body [87].

An example of a battery capable of providing electrical power for interactive electronic textiles was recently developed by a German research team led by The Fraunhofer Institute for Reliability and Micro-integration (FhG-IZM, Berlin, Germany). This research team developed a small battery that can be printed on a substrate and fabricated at high production speeds in button-sized or coin-type format at a cost below one USD. The battery is fabricated by screen printing a thick layer of a silver-oxide based paste, then applying a thin sealing layer. The final result is a textile substrate with a printed 120 μm thick AgO-ZN battery. These batteries can be printed on a variety of substrates [88].

2.6. Planar Fashionable Circuit Board (P-FCB)

The P-FCB is one of the new technologies allowing implementation of a circuit board on a plain fabric patch for wearable electronics applications. It features a soft and flexible impression just as normal clothes.

The P-FCB substrate is fabricated using woven fabrics. The planar electrodes are deposited on the fabric patch directly by silk screening of conducting epoxy or by gold sputtering. First, the circuit board is silkscreen printed on the fabric patch. Then the IC is placed on the fabric and wire-bonded to the patterned electrodes. Finally, the IC is molded with non-conductive epoxy (Figure 16a) [89].

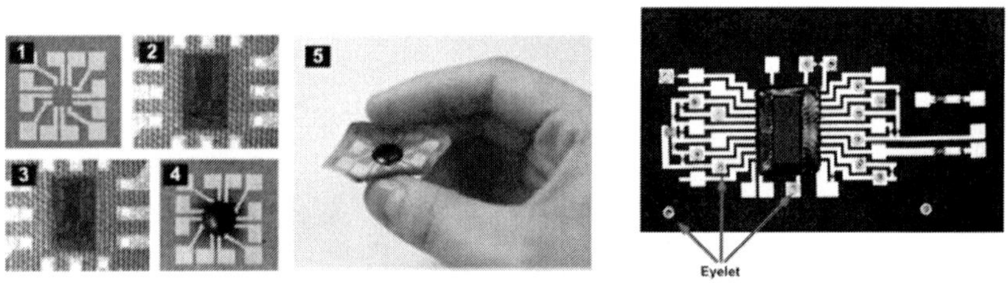

(a) (b)

Figure 16. (**a**) Single-layered P-FCB system manufacturing process; (**b**) Implementation of multi-layer connection using eyelet.

With these techniques, KAIST researchers also developed a multilayer circuit (Figure 16b). The previous research highlight how electrical and mechanical characteristics and valuable system design parameters are obtained, such as maximum power consumption, maximum current density, crosstalk between neighboring two lines, and durability [90].

2.7. Wearable Antenna

Thanks to the rapid progress on the fabrication of conductive textiles, a significant development of wearable antennas has begun, exploiting new flexible and conformable smart structures [91].

An antenna is essential, if the purpose is to develop a wearable and autonomous system. It allows one to transfer information from the sensors hosted inside the garment to a control unit or to monitor other electronic parameters.

A wearable antenna is thus the bond that integrates clothes into the communication system, making electronic devices less obtrusive. To achieve good results, wearable antennas have to be thin, lightweight, low maintenance, robust, inexpensive and easily integrated in radio frequency (RF) circuits. Planar structures, flexible conductive and dielectric materials are specific requirements for wearable antennas [92,93].

Several properties of the materials influence the behavior of the antenna. For instance, the bandwidth and the efficiency of a planar microstrip antenna are mainly determined by the permittivity and the thickness of the substrate [94,95]

In general, textiles present a very low dielectric constant that reduces the surface wave losses and increases the impedance bandwidth of the antenna. Therefore it is important to know how these characteristics influence the behavior of the antenna in order to minimize unwanted effects.

Another issue regards the movement of the body that can deform the spatial geometry of the antenna and affect its performance. When the textile fabric adapts to the surface topology it bends and deforms, causing changes to its electromagnetic properties and thus influencing the antenna performance [96]. Thus, a wearable FM antenna should be designed so as to be wider than the FM broadcast band (about 81–130 MHz) not to suffer from the detuning caused by the human body [97].

To summarize, the guidelines for a correct antenna project are shown below:

- choosing the correct positioning of the textile antenna [98,99];
- the textile antenna must be made with an accurate thickness stacking the different fabrics [99];
- the geometrical dimensions of the patch must remain stable [100];
- the connections between the layers must not affect the electrical properties and the connections with other parts of "e-garments" have to be stable and robust [101–103].

If some of these points are not followed, undesired effects may occur to the functioning of the device.

Related Works

Researchers at Katholieke Universiteit Leuven and Universiti Malaysia Perlis were the first to develop a fully textile waveguide antenna using a material inspired unit cell that is also used in composite right/left-handed transmission lines. The antenna is compact, robust and can be used for 2.45 and 5.4 GHz dual-band WLAN applications [104].

Patria (Halli, Finland) is a company with expertise in textile antenna design. It develops textile antennas composed by conventional or industrial fabrics, and typically conductive antenna parts are made out of modern conductive fibers [105].

Textiles RFID is a particular solution of antenna. In this sector TexTrace AG (Frick, Switzerland) provides the manufacturing line as well as the components for industrial in-house production of woven RFID labels. Integrate RFID and the label will provide added value from garment manufacturing through logistics to sales and after-sales management [106].

2.8. Stretchable Interconnection

Deformable electronic circuits, in particular in bio-medical application, are needed. E-textiles are the candidate technology for this scope but an important aspect regards the interconnections between components and devices. In most cases the approach used is to develop sinuous electroplated metallic wires in a stretchable substrate material [107].

Figure 17 presents a stretchable interconnection structure. This configuration, called "horseshoe-shape", is able to accommodate large deformation in response to a mechanical stress, preserving the electrical properties [108].

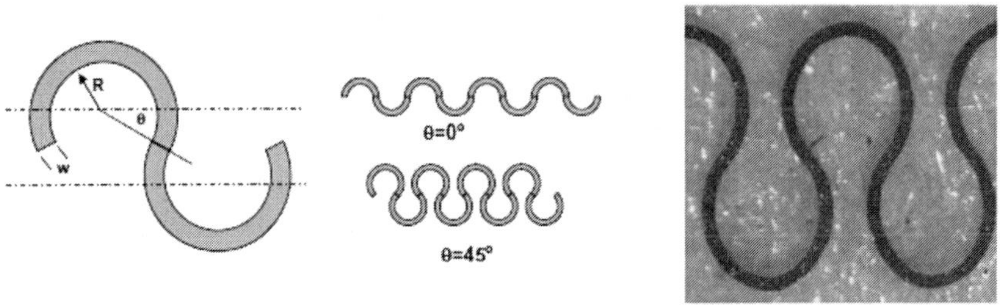

(a) (b)

Figure 17. (**a**) Notable measures of a horseshoe design: inner radius (R), joining angle (θ) and width of the metal track (w); (**b**) Horseshoe metal interconnects embedded into a Sylgard 186 PDMS matrix.

Indeed, with high-frequency signal the interconnection have to be stretchable and therefore the substrate and the conductors have to be stretchable as well. Usually, a polymeric material is chosen as encapsulating substrate, because it can be stretched and also because it may be bio-compatible [109].

Interconnections made with these techniques coupled with conductive yarns can go beyond actual issues regarding the lack of robustness in e-textiles circuit.

3. DISCUSSION

The methods presented in the previous section are quite different and each one has very specific feature such as conductivity, flexibility, biocompatibility, mechanical resistance and washability. However, only a few approaches are able to satisfy all these requirements at the same time. The section below has the aim to clarify some features about the materials and methods.

3.1. Conductive Wires

Especially for clothing, tactile properties such as stretch, recovery, drape, shear and handle are quite important. For this reason the fibers that are used should be fine and fabrics should have a low weight per unit area (not more than 300 g/m^2). These demands are inconsistent with the materials and geometries that are needed for a reasonable electrical conductivity, because the incorporation of elements such as metal wires within textiles increases stiffness and reduces elasticity [110].

Often in literature no distinction is made between metal wires and metal fibers. However, the company Sprint Metal (Hemer, Germany) defined metal fibers and wires according to their diameter. While a fine wire has a diameter between 30 μm to 1.4 mm, a metal fiber possesses a diameter of 2 to 40 μm [111].

The advantages of metal fibers are their strength, composition, biological inertness and ready availability in textile form at low costs. Due to its inertness it is not sensitive to washing or sweating.

However, they cannot provide uniform heating and their brittle characteristics can damage spinning machinery over time. Additionally, they are heavier than most textile fibers making homogeneous blends difficult to produce [112].

Using the 'conductive thread' approach, no additional step after manufacturing of the fabric is required to establish conductivity. The conductivity of these conductive threads lies in the range of 10–500 Ω/m [113].

For instance, Figure 4a shows the copper wire twisted around the yarn and their locations are not precisely defined within the fabric because of their helical path around the yarn [114]. The conductance of a wire in the fabric features about 17.8 Ω/m.

3.2. Treated Coatings

The advantage of coatings is that they are suitable for many fibers types and they produce good conductivity without significantly altering existing key substrate properties such as density, flexibility and handle. Nevertheless adhesion between the metal and the fibers as well as corrosion resistance can lead to problems.

It is possible to make a conductive fabric through a coating treatment even though the main applications for these textiles are protection against static electricity charge and electromagnetic interference (EMI). The surface resistivity (Rs) of normal textiles goes from more than 10^5 Ω/sq down to antistatically finished textiles and conductive textiles for EMI and heating applications, where Rs can be lower than 100 Ω/sq. It is usual to incorporate or impregnate organic antistatic agents [98].

At present the most important ways of producing conductive fibers are:

- incorporation of conductive fillers (carbon black, metal wires, graphite and metal powder or flakes of Al, Cu, Ni, Ag).
- blended fabrics using conductive metallic or polymeric fibers (such as polypyrrole, polyaniline) (Figures 16b and 18a) [115].

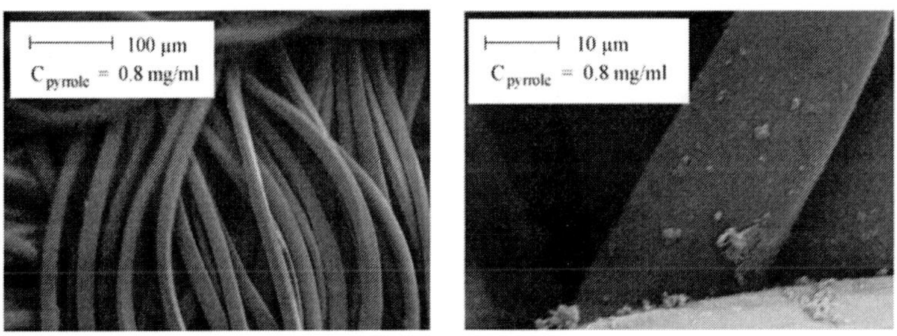

(a) (b)

Figure 18. SEM image of the polypyrrole coated textile ($C_{pyrrole}$ = 0.8 mg/mL), (**a**) 100 μm scale and (**b**) 10 μm scale.

However, the surface resistance of polypyrrole-coated ($C_{pyrrole}$) polyester fabrics is another method to functionalize the textiles. Rs decreased marginally from 106 Ω/sq to 103 Ω/sq with increasing pyrrole concentration within 0.2 mg/mL until the concentration reached a value of about 0.4 mg/mL, above which the rate of decrease diminished. The decrease in surface resistance is due to the increase in the thickness of the conducting polymer layer [116].

The main advantage of functionalized yarns is that the textile properties of the yarns are preserved, in particular they remain flexible.

3.3. Conductive Fabric PETEX

Normally, a fabric made with conductive fibers has a structure according to Figure 19. If the aim is to exploit the conductive fibers to create an electrical circuit, we must take into account the irregular shape of the fabric. It is not possible to obtain a geometric repeatability of the position of the contact points between the conductive fibers.

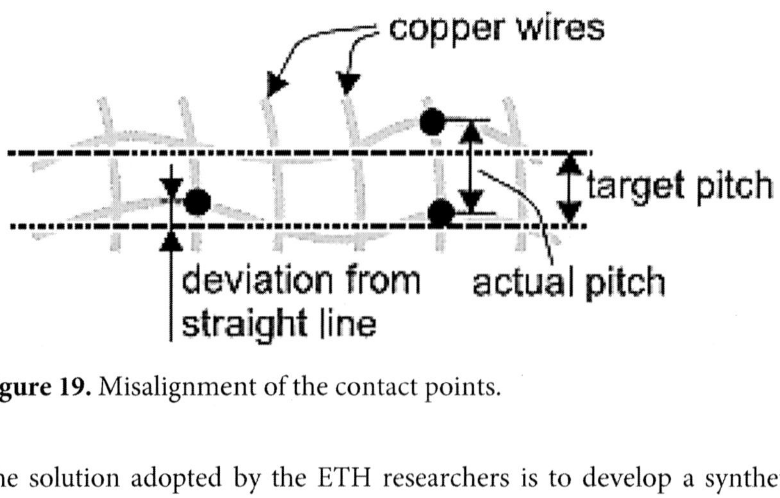

Figure 19. Misalignment of the contact points.

The solution adopted by the ETH researchers is to develop a synthetic fabric and it consists of woven polyester monofilament yarn (PET) with diameter 42 μm and copper alloy wires with diameter 50 ± 8 μm (AWG461). Each copper wire itself is coated with a polyurethane varnish as electrical insulation. The wires in warp direction possess a higher tensile strength compared to the wires in weft direction. These differences support a reliable manufacturing process and minimize breakage.

Figures 18b and 20a below, show the differences between a normal electronic textile structure and PETEX [47].

(a) (b)

Figure 20. SEM image of fabric with copper fiber (**a**) and PETEX (**b**).

3.4. MIT CAD Embroidery

This technique allows precisely specifying the circuit layout and stitching pattern in a computer-aided design (CAD) environment, from which any number of articles can be sewn under machine control. This process also allows control and integration of yarns with different electrical properties, for instance, different resistances. Embroidery offers advantages over knitting or weaving. Conductive thread and yarn embroidery can be accomplished on single or multiple layers of fabric or can be applied on various types of textile and apparel products in one step [39].

3.5. Conductive Inks

Direct printing of conductive tracks on a fabric is a versatile technique but sometimes shows its limits. Most conductive inks and pastes are based on silver filler and suffer from brittleness. During printing, one of the limits of the silver paste is the thickness. For instance, several passes are needed to achieve a layer thick 40 μm and the electrical resistance depends on the number of passes applied (Figure 21).

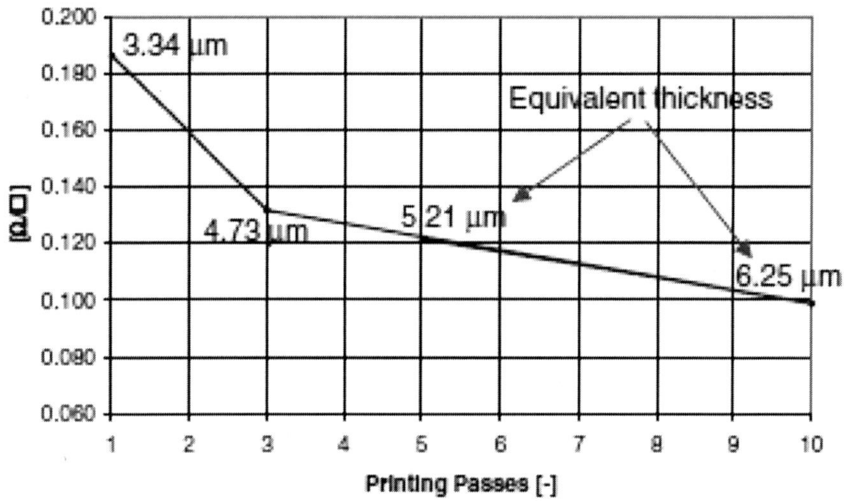

Figure 21. Sheet resistance vs. printing passes.

Another problem is the mechanical resistance. If you create a condition as shown in Figure 22a, the risk is to have a break in the conductivity. In particular the electrical conductivity changes both for the number of passes and for the number of creasing interactions (Figure 22b) [47].

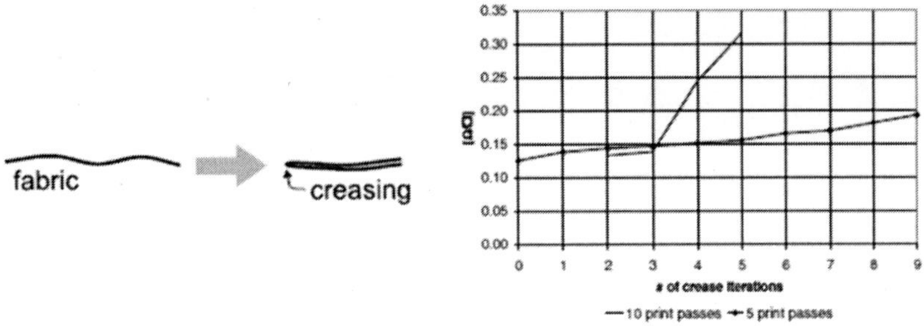

(a) (b)

Figure 22. (**a**) Creasing of the fabric with printed transmission lines; (**b**) DC resistance changes vs. number of creasing iterations.

The researchers of North Carolina Institute, developed a new method to control the durability of the printed circuits, which includes coating of the printed lines. The viscosity of the ink dictated the performance of the printed media during washing trials.

The results showed that higher viscosity inks stay on the surface, resulting in better short-term results. (Figure 23) The ink used was produced by Creative Materials and showed rather a smooth and lower impedance profile than other commercial conductive inks. This ink is based on silver particles and the lines printed on Evolon® show 42 times higher resistivity than standard PCBs (0.50 mΩ/m). However, Evolon® like structures with the highest viscosity ink represent the best solution in terms of durability and conductivity also after washing. This substrate-ink combination protected by a TPU shows the best compromise to become a low-cost electronic textile solution for wearable computing interconnects [117].

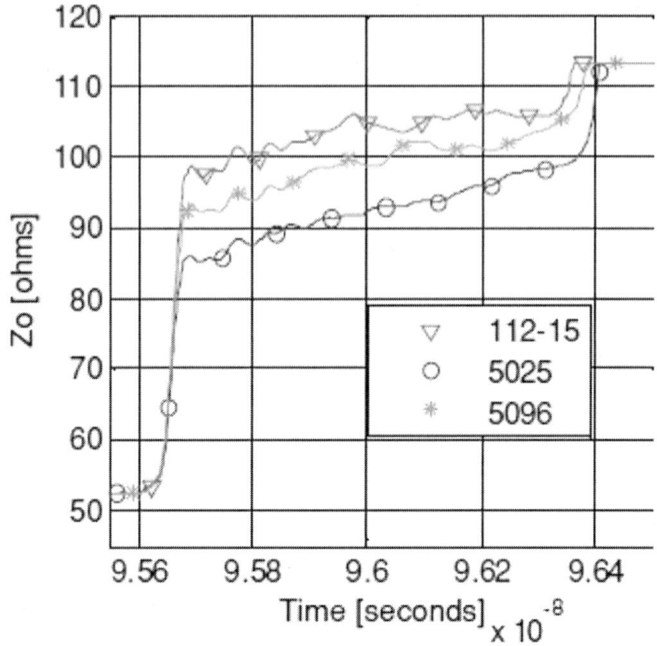

Figure 23. Evolon impedence profile before five washes. The three lines show the comparison between different conductive inks (Dupont 5025, Dupont 5096, Creative Materials 112-15).

3.6. Stretchable Sensors

Usually, textile materials composed of fibers form complex networks of conducting parts that make multiple contacts. During deformation a number of mechanisms take place:

- the number of contact points changes;
- fibers are extended;
- fiber cross-section is decreased.

These factors listed above may compromise the stretch sensors qualities [118]. Polymers are interesting when developing strain sensors with large strain [1]. The stretch sensor developed by Mattmann et al. [119] consists of a mixture of 50 wt-% thermoplastic elastomer (TPE) and 50 wt-% carbon black particles and is fiber-shaped with a diameter of 0.315 mm.

This sensor configuration was characterized using a strain tester and measuring the resistance (extension-retraction cycles): it showed a linear resistance response to strain, a small hysteresis, no ageing effects and a small dependence on the strain velocity. The attachment of the sensor to the textile is realized using a silicone film.

The graphs (Figure 24) below show the behavior of the sensor and the typical resistance vs. time plot. The sensor was cycled between 0% and 80% strain at a speed of 200 mm/min and waiting times at minimal and maximal strain of 2 min. In the lower plot the measured resistance is shown which varying between 2 kΩ and 19 kΩ. Instead, Figure 25 shows a linear rise in resistance when applying strain and only a small hysteresis.

Commercially available strain gauges reach a higher than 80% linearity but at a reduced working range of less than 1%. The sensor presented above has a high sensitivity of 1.25 kΩ/mm with a stable sensor properties while in continuous usage. Furthermore the sensor is washable.

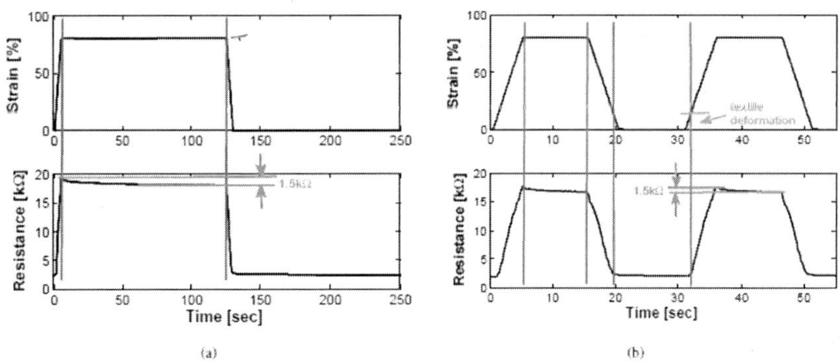

Figure 24. **(a)** Typical response of sensor to a given strain (sensor length 2 cm) waiting time 2 min; **(b)** waiting time 10 s.

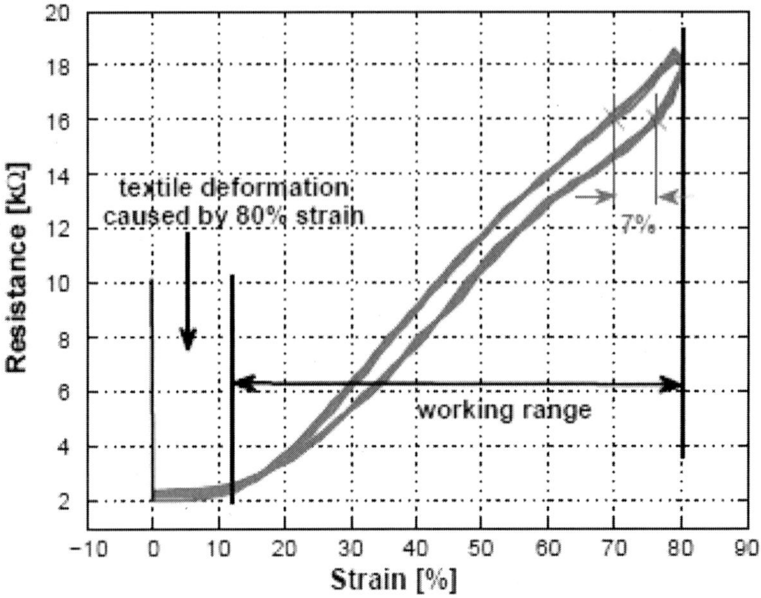

Figure 25. Typical response of sensor to a given strain.

A similar approach is that used by Shyr et al., in which they developed three different geometries of a strain-resistance sensors using Lycra fiber and carbon coated polymide fiber: flat, tubular and belt webbings. It was found that tensile hysteresis and contact resistance significantly influence the tensile elasticity and the resistance sensitivity of the webbings, respectively. Their results showed that all the webbings had a good linear elasticity relationship between tensile load and strain when stretched with the aim to reduce the tensile hysteresis, decreasing the friction among the conductive yarns [118]. The stability of resistance sensitivity and the elasticity are two relevant properties for e-textiles when used as a strain-resistance sensor.

3.7. Pressure Sensors

The pressure sensor developed by Meyer et al. [9] has a spatial resolution of 2 × 2 cm and an average error below 4% within the measurement range 0 to 10 N/cm^2. It is a pressure sensor based on capacitive behavior. The measured capacitance of a single electrode of the prototype with the textile spacer varies from 3.5 pF uncompressed to 5.8 pF at a pressure of 5 N/cm^2 for the textile spacer with 6 mm thickness. Figure 26 shows the capacitance vs. pressure with a hysteresis up to 30%. The high value of hysteresis is due to the dielectric material interposed between the two conductive faces capacitor.

An ideal pressure sensor must ensure repeatable measures over time with a low hysteresis value. Therefore, the textile pressure sensors are applicable in areas where high accuracy is not required.

3.8. Electrochemical Sensors

One of the common disadvantages of chemical/biochemical sensors is the need to calibrate the response from time to time due to either small changes in the activity of the sensor itself and/or the contamination of the sensor surface during operation. This has seriously limited the application of these sensors in continuous monitoring scenarios for long time. The major obstacles still remain, preventing the realization of practical, robust wearable sensors suitable for routine use by the general population.

Among these limitations, we have the lack of appropriate methods to seamlessly integrate electronics and wireless transmitters into the sensor package. The graph below (Figure 27) shows the performance of the wearable sweat sensor during an exercise trial. Reference pH measurements are taken along with heart rate and breathing rate data. The analysis were taken by placing the electrode on a reference fabric patch, as performed in previous studies [120].

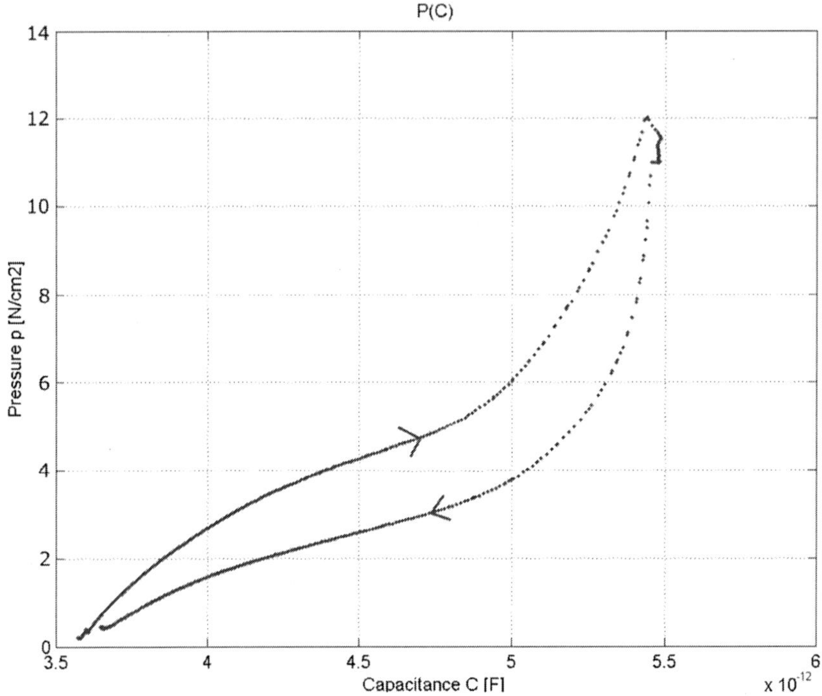

Figure 26. Hysteresis of a singular pressure electrode.

3.9. Power Supply

In the following Table 4, an overview of different possible energy sources and the amount of energy that they can generate is given [120,121].

From these data, we can assume that the energy harvesting coming from our body is very small although it is always present. If the wearable technologies will take advantage of these energy sources, they must feature a low electrical consumption and a high efficiency harvesting system.

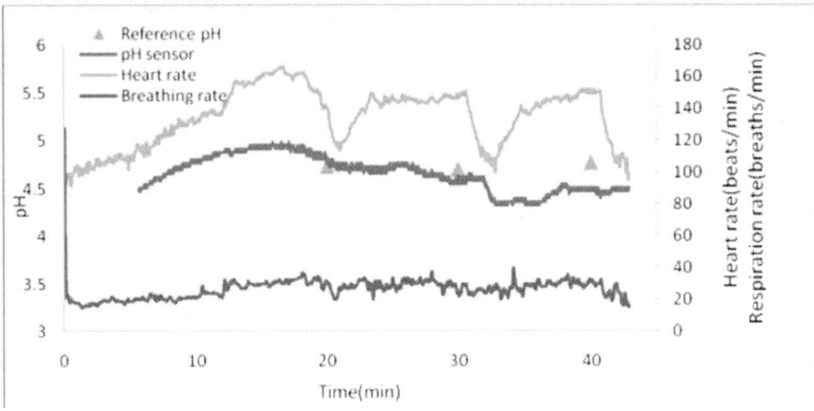

Figure 27. Wearable sweat sensors performance.

Table 4. Comparison of different energy source and their energy capability.

Energy Source	Available Amount of Energy	Remarks
Primary batteries (Li)	400 Wh/kg, 200 Wh/L	
Secondary batteries (Li-Ion)	75 Wh/kg, 200 Wh/L	Lifetime: 2000 cycles
Si solar cells	20 W/m^2	Light source needed
Recovery of body heat	0.01 W/m^2	
Power harvesting from breathing	0.4 W	Uncomfortable for wearer
Power harvesting from walking	0.25 W	Continuous walking necessary
Microcombustion	10–50 W/cm^3	Fuel needed

3.10. Planar Fashionable Circuit Board (P-FCB)

The measured sheet resistance of the conducting film on the P-FCB is 44 mΩ/m, as determined by the four-point probe method. Although no noticeable degradation has been observed, even after 50 laundry cycles, more durability examinations will be required with an automatic testing system.

The resistance measurement results of the proposed via and conductive adhesive and bandwidth of a P-FCB transmission line 15 cm long and 1mm wide are shown in Figures 26b and 28a. Its bandwidth is 80 MHz, and this is sufficient to deal with bio-signal processing. The analysis were performed with a vector network analyzer [122]. To get an average value, as many as 100 connections were formed and measured. The resistances of the proposed via and conductive adhesive measured 0.24 W and 0.34 W, respectively [123].

Mechanical strength of the P-FCB was tested through washing tests; after 50+ cycles, the electrical characteristics change was negligible [123]. As a result, the monitoring shirt can be easily made at low cost, and the wearability and flexibility are improved.

3.11. Wearable Antenna

Antennas for wearable applications need to be comfortable to wear and traditional technology with rigid substrates is not well suited to this.

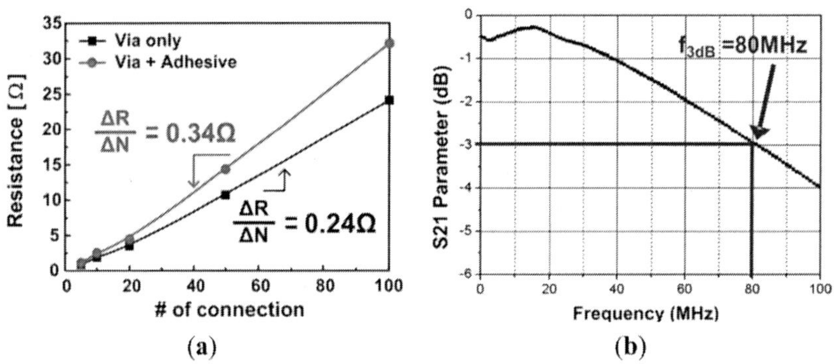

Figure 28. (**a**) Measurement result of the proposed via and conductive adhesive resistance; (**b**) Bandwidth of a P-FCB transmission line (15 cm long, 1 mm wide).

In this sector the dielectric properties of the material that make up the antenna are more relevant. A wrong choice of material would compromise the performance of the device.

The constitutive parameter of dielectrics is the permittivity, ε, a complex value parameter and it usually expressed by Equation (1) below:

$$\varepsilon_r: \quad \varepsilon = \varepsilon_0 \cdot \varepsilon_r = \varepsilon_0 \cdot (\varepsilon_r' - \mathrm{j} \cdot \varepsilon_r'') \tag{1}$$

where ε_0 is the permittivity of vacuum, 8.854×10^{-12} F/m [124]. The dielectric properties are a function of different physical parameters: temperature, frequency, surface roughness, moisture content, purity and homogeneity of the material.

Table 5 below shows the dielectric constant $\boldsymbol{\varepsilon}'_r$ and the ratio of the imaginary part to the real part, also known as $\tan \delta = \boldsymbol{\varepsilon}''_r / \boldsymbol{\varepsilon}'r$ [94].

The low dielectric constant (between 1 and 2) reduces the surface wave losses that are tied to guided wave propagation within the substrate. Therefore, lowering the dielectric constant increases spatial waves and hence increases the impedance bandwidth of the antenna [125].

Table 5. Dielectric Properties of normal fabrics.

Nonconductive Fabric	ε'_r	tan δ
Condura ®	1.90	0.0098
Cotton	1.60	0.0400
Polyester	1.90	0.0045
Quartzel ®	1.95	0.0004
Cordura/Lycra®	1.50	0.0093

In addition, the moisture can affect the dielectric behavior of textiles. In particular, when water is absorbed by the fibres, an increase of the dielectric constant and loss tangent occurs. A textile cover or superficial treatment may then provide a sufficient protection from the humidity and varying climatic conditions [100].

Researchers of Tampere University of Technology, in collaboration with Patria, developed a robust textile antenna able to operate in different environmental conditions. They demonstrated that different functional textile layers can improve and optimize the antenna performance. In this occasion they succeeded to protect the antenna against moisture and others environmental factors [126].

The thickness of the dielectric material changes the bandwidth of the planar antenna. It is necessary to find a compromise between efficiency and bandwidth, well as to influences the geometric sizing of the antenna [125].

For the antenna design, the relevant parameter is the conductivity of the fabric (σ), which unit is Siemens per meter (S/m). It is expressed by this Equation (2):

$$\sigma = 1/(\rho_s \cdot t) \tag{2}$$

where ρ_S is the surface resistivity and t is the thickness of the fabric. An issue regarding the conducibility of conductive fabrics for antenna is the homogeneity of σ value; In fact, there are several discontinuities inside the fabric and these affect the current flow [127].

Another aspect that influences the conductivity is the structure of the fabric. The current flow depends on the type of textile structure used and to minimize the conductive losses it is better to align the conductive paths with the current direction [98].

Finally, a correct choice of materials combined with textile structures can improve the performance of the textile antenna equaling the features guaranteed by traditional antennas.

3.12. Stretchable Interconnections

Stretchability of the electrical conductors is a requirement that can be satisfied with different designs. The design optimization is done using FEM modeling (Figure 29) and other tests. Gonzalez et al. evaluated the “horseshoe-shape” to be the optimal solution for this purpose [108]. In this way the damage in the conductor is significantly decreased with a larger deformability. They demonstrated that in case of multiple and narrow metallisations the stretchability of the circuit reaches 100% of the deformability.

Hu et al. used a different approach to obtain a stretchable conductive textile. With a simple “dipping and drying” process using single-walled carbon nanotube (SWNT) ink, they produced highly conductive textiles with conductivities of 125 $S \cdot cm^{-1}$ and sheet resistances less than 1 Ω/sq. In this way, such conductive textiles preserve the same stretchability of the normal fabric. Indeed, the porous structure of the textiles facilitates accessibility from any electrolyte and such porous and stretchable conductors find a wide range of applications in e-textiles field [128].

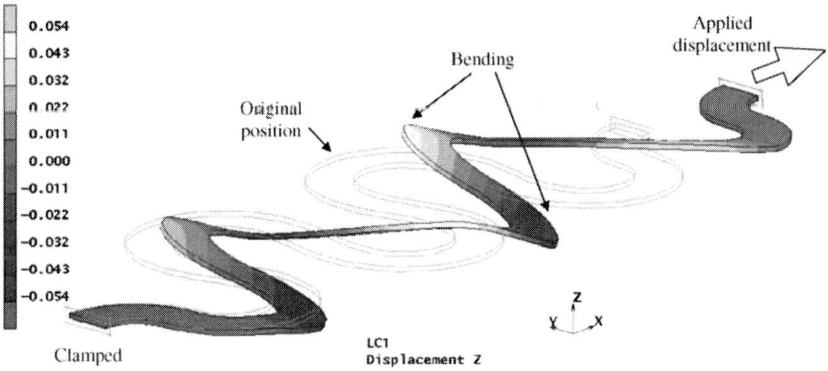

Figure 29. FEM simulation of the “horseshoe-shape”.

4. CONCLUSION/OUTLOOK

Textiles represent an attractive class of substrates for realizing wearable bio-sensors. Electronic textiles, or smart textiles, describe the convergence of electronics and textiles into fabrics which are able to sense, compute, communicate and actuate. As many different electronic systems can be connected to any clothing, a wearable system becomes more versatile, and the user can change its look depending on environmental changes and individual preference.

The vision of wearable computing describes future electronic systems as an integral part of our everyday clothing serving as intelligent personal assistants. Therefore, such wearable sensors must maintain their sensing capabilities under the demands of normal wear, which can impose severe mechanical deformation of the underlying garment/substrate.

One promising approach to reduce the rigidity of electronic textiles and enhance its wearability is to replace PCBs by flexible electronics.

In this review we wanted to explain how it is possible to develop a Smart Textile. Some methods show advantages with respect to others, but in our opinion and in according to the consulting company Smart Garment People (Lancashire, UK), while some manufacturers are very experienced with electronics and others with textiles, very few do both well.

Current advances in textile technologies, new materials, nanotechnology and miniaturized electronics are making wearable systems more feasible but the final key factor for user acceptance of wearable devices is the fit comfort. We are convinced that this goal can only be achieved by addressing mechanical resistance, and durability of the materials in what is recognized to be a harsh environment for electronics: the human body and society.

Finally, we consider relevant that the development of smart textiles requires a multidisciplinary approach in which knowledge of circuit design, smart materials, micro-electronics and chemistry are fundamentally integrated with a deep understanding of textile fabrication.

AUTHOR CONTRIBUTIONS

Authors contributed equally to the manuscript in its final form.

REFERENCES

1. Langereis, G.R.; Bouwstra, S.; Chen, W. Sensors, Actuators and Computing Architecture Systems for Smart Textiles. In Smart Textiles for Protection; Chapman, R., Ed.; Woodhead Publishing: Cambridge, UK, 2012; Volume 1, pp. 190–213.

2. Custodio, V.; Herrera, F.J.; López, G.; Moreno, J.I. A review on architectures and communications technologies for wearable health-monitoring systems. Sensors 2012, 12, 13907–13946.

3. Coosemans, J.; Hermans, B.; Puers, R. Integrating wireless ECG monitoring in textiles. Sens. Actuators A Phys. 2006, 130–131, 48–53.

4. Linz, T.; Gourmelon, L.; Langereis, G. Contactless EMG sensors embroidered onto textile. Proceedings of the 4th International Workshop on Wearable and Implantable Body Sensor Networks, Aachen, Germany, 26–28 March 2007; Volume 13, pp. 29–34.

5. Löfhede, J.; Seoane, F. Thordstein, Soft textile electrodes for EEG monitoring. Proceedings of 2010 the 10th IEEE International Conference on Information Technology and Applications in Biomedicine (ITAB), Corfu, Greece, 2–5 November 2010; pp. 1–4.

6. Löfhede, J.; Seoane, F.; Thordstein, M. Textile electrodes for EEG recording—A pilot study. Sensors 2012, 12, 16907–16919.

7. Sibinski, M.; Jakubowska, M.; Sloma, M. Flexible temperature sensors on fibers. Sensors 2010, 10, 7934–7946.

8. Omenetto, F.; Kaplan, D.; Amsden, J.; Dal Negro, L. Silk Based Biophotonic Sensors. Patent US 2013/0330710, 2013.

9. Meyer, J.; Lukowicz, P.; Tröster, G. Textile Pressure Sensor for Muscle Activity and Motion Detection. Proceeding of the 10th

IEEE International Symposium on Wearable Computers, Montreux, Switzerland, 11–14 October 2006.

10. Coyle, S.; Lau, K.-T.; Moyna, N.; O'Gorman, D.; Diamond, D.; Di Francesco, F.; Costanzo, D.; Salvo, P.; Trivella, M.G.; De Rossi, D.E.; et al. BIOTEX—Biosensing textiles for personalised healthcare management. IEEE Trans. Inf. Technol. Biomed. 2010, 14, 364–370.

11. Zadeh, E. Flexible biochemical sensor array for laboratory-on-chip applications. Proceeding of the International Workshop on Computer Architecture for Machine Perception and Sensing, Montreal, QC, Canada, 18–20 September 2006; pp. 65–66.

12. Vatansever, D.; Siores, E.; Hadimani, R.; Shah, T. Smart Woven Fabrics in Renewable Energy Generation. In Advances in Modern Woven Fabrics Technology; Vassiliadis, S., Ed.; InTech: Rijeka, Croatia, 2011; pp. 23–38.

13. Baurley, S. Interactive and experiential design in smart textile products and applications. Pers. Ubiquitous Comput. 2004, 8, 274–281.

14. Black, S. Trends in Smart Medical Textiles. In Smart Textiles for Medicine and Healthcare: Materials, Systems and Applications; Van Langenhove, L., Ed.; University of Ghent: Ghent, Belgium, 2007; Volume 1, pp. 10–22.

15. Edmison, J.; Jones, M.; Nakad, Z.; Martin, T. Using piezoelectric materials for wearable electronic textiles. Proceedings of the 6th International Symposium on Wearable Computers (ISWC), Seattle, WA, USA, 7–10 October 2002; pp. 41–48.

16. Bedeloglu, A.; Demir, A.; Bozkurt, Y.; Sariciftci, N.S. A Photovoltaic Fiber Design for Smart Textiles. Text. Res. J. 2009, 80, 1065–1074.

17. How to Get What You Want? Available online: http://www.kobakant.at/DIY/?p=48 (accessed on 11 January 2014).

18. Pacelli, M.; Loriga, G.; Taccini, N.; Paradiso, R. Sensing Fabrics for Monitoring Physiological and Biomechanical Variables: E-textile solutions. Proceeding of the IEEE/EMBS International Summer School on Medical Devices and Biosensors, St. Catharine's College, Cambridge, UK, 19–22 August 2007; pp. 1–4.

19. Dias, T. Development and Analysis of Novel Electroluminescent Yarns and Fabrics for Localised Automotive Interior Illumination: El Yarns and Fabrics. Text. Res. J. 2012, 82, 1164–1176.

20. Janietz, S.; Gruber, B.; Schattauer, S.; Schulze, K. Integration of OLEDs in Textiles. Adv. Sci. Technol. 2012, 80, 14–21. [G]

21. Munro, B.J.; Steele, J.R.; Campbell, T.E.; Wallace, G.G. Wearable Textile Biofeedback Systems: Are They too Intelligent for the Wearer? In Wearable eHealth Systems for Personalised Health Management: State of the Art and Future Challenges; Lymberis, A., De Rossi, D., Eds.; IOS Press—STM Publishing House: Amsterdam, The Netherlands, 2005; Volume 108, pp. 271–277.

22. Salonen, P.; Hurme, L. A novel fabric WLAN antenna for wearable applications. Proceeding of IEEE International Symposium on Antennas and Propagation Society, Columbus, OH, USA, 22–27 June 2003; Volume 2, pp. 700–703.

23. Popular Embroidery Techniques Used to Decorate Fabrics. Available online: http://nanetteparker.hubpages.com/hub/Popular-Embroidery-Techniques-Used-to-Decorate-Fabrics (accessed on 4 February 2014).

24. Creative Sewing. Available online: http://www.creativesewing.co.nz/ (accessed on 4 February 2014).

25. Loominous. Available online: http://www.loominous.co. uk/studio.html (accessed on 4 February 2014).

26. Cornell University—Fabrics of Our Livelihoods. Available online: http://smallfarms.cornell.edu/2011/07/04/fabrics-of-our-livelihoods/ (accessed on 4 February 2014).

27. CMI. Available online: https://www.colonialmills. com/PublicStore/catalog/Braiding-Process,156.aspx (accessed on 4 February 2014).

28. Textile Innovation Knowledge Platform. Available online: http://www.tikp.co.uk/knowledge/technology/coating-and-laminating/ laminating (accessed on 4 February 2014).

29. Custom Fabric Printing. Available online: http://sophiasdecor.blogspot.it/2012/09/inside-spoonflower-custom-fabric.html (accessed on 4 February 2014).

30. Durable water repellent. Available online: http://en. wikipedia.org/wiki/Durable_water_repellent (accessed on 4 February 2014).

31. Post, E.R.; Orth, M.; Russo, P.R.; Gershenfeld, N. E-broidery: Design and fabrication of textile-based computing. IBM Syst. J. 2000, 39, 840–860.

32. Systenx Project. Available online: http://www.systex.org/projects (accessed on 10 September 2013).

33. Proetex Project. Available online: http://www. proetex.org/related_projects.htm (accessed on 10 September 2013).

34. Resistat Fiber Collection. Available online: http://www. resistat.com/ (accessed on 12 January 2014).

35. Redström, M.; Redström, J.; Mazé, R. IT + Textiles, 1st ed.; The Interactive Institute: Boràs, Sweden, 2005; pp. 59–93.

36. Sophitex Ltd. Available online: http://www.sophitex.com (accessed on 31 January 2014).

37. LessEMF. Available online: http://www.lessemf.com/fabric.html (accessed on 15 July 2013).

38. McFarland, E.G.; Carr, W.W.; Sarma, D.S.; Dorrity, J.L. Effects of Moisture and Fiber Type on Infrared Absorption of Fabrics. Text. Res. J. 1999, 69, 607–615.

39. Mac, T.; Houis, S.; Gries, T. Metal Fibers. Proceeding of the International Conference on Shape Memory and Superelastic Technologies, Baden-Baden, Germany, 3–7 October 2004; Volume 47.

40. Elektrisola Feindraht AG. Available online: www.textile-wire.com (accessed on 11 January 2014).

41. Spoerry & Co AG, Swiss Shield®. Available online: www.swiss-shield.ch (accessed on 12 January 2014).

42. Gimpel, S.; Moehring, U.; Mueller, H.; Neudeck, A.; Scheibner, W. The galvanic and electrochemical modification of textiles. Band-und Flechtind. 2003, 40, 115–120.

43. Abouraddy, A.F.; Bayindir, M.; Benoit, G.; Hart, S.D.; Kuriki, K.; Orf, N.; Shapira, O.; Sorin, F.; Temelkuran, B.; Fink, Y. Towards multimaterial multifunctional fibres that see, hear, sense and communicate. Nat. Mater. 2007, 6, 336–347.

44. Hamedi, M.; Forchheimer, R.; Inganäs, O. Towards woven logic from organic electronic fibres. Nature Mater. 2007, 6, 357–362.

45. Müller, C.; Hamedi, M.; Karlsson, R.; Jansson, R.; Marcilla, R.; Hedhammar, M.; Inganäs, O. Woven electrochemical transistors on silk fibers. Adv. Mater. 2011, 6, 898–901.

46. Scheibner, W.; Feustel, M.; Moehring, U.; Hofmann, J.; Linz, T. Textile electrical connecting cables. Band-und Flechtind. 2003, 40, 76–78.

47. Locher, I. Technologies for System-on-Textile Integration. Ph.D. Thesis, Swiss Federal Institute of Technology (ETH), Zurich, Switzerland, 2006.

48. Baltex. Available online: http://www.baltex.co.uk (accessed on 11 November 2013).

49. Thermshield. Available online: http://www. thermshield. com/ ThermshieldPages/MetallizedFabric.html (accessed on 12 January 2014).

50. Ohmatex, Smart Textile Technology. Available online: http:// www.ohmatex.dk (accessed on 3 June 2013).

51. Parashkov, R.E.; Becker, E. Large area electronics using printing methods. Proc. IEEE 2005, 93, 1321–1329.

52. Blayo, A.; Pineaux, B. Printing processes and their potential for RFID printing. Proceeding of Joint sOc-UESAI Conference, Grenoble, France, 7–10 October 2005; pp. 27–30.

53. Stassi, S.; Cauda, V.; Canavese, G.; Manfredi, D.; Roppolo, I.; Martino, P.; Chiolerio, A. Nanosized Gold and Silver Spherical, Spiky, and Multi-Branched Particles. In Handbook of Nanomaterials Properties; Bhushan, B., Ed.; Springer-Verlag: Berlin/Heidelberg, Germany, 2014; pp. 199–202.

54. Tiberto, P.; Barrera, G.; Celegato, F.; Coisson, M.; Chiolerio, A.; Martino, P.; Pandolfi, P.; Allia, P. Magnetic properties of jet-

printer inks containing dispersed magnetite nanoparticles. Eur. Phys. J. B. 2013, 83, 173–179.

55. Lee, H.-H.; Chou, K.-S.; Huang, K.-C. Inkjet printing of nanosized silver colloids. Nanotechnology 2005, 16, 2436–2441.

56. Camarchia, V.; Chiolerio, A.; Cotto, M.; Fang, J.; Ghione, G.; Pandolfi, P.; Pirola, M.; Pirola, M.; Quaglia, R.; Ramella, C. Demonstration of inkjet-printed silver nanoparticle microstrip lines on alumina for RF power modules. Org. Electron. 2014, 15, 91–98.

57. Printing, S.; Sauer, M.; Meilchen, S.; Kalleder, A.; Mennig, M.; Schmidt, H. Screen printing. In Sol-Gel Technologies for Glass Producers and Users; Michel, A., Aegerter, A., Mennig, M., Eds.; Springer: Norwell, MA, USA, 2004; Volume 1.

58. Numakura, D. Advanced Screen Printing—Practical Approaches for Printable & Flexible Electronics; DKN Research: Haverhill, MA, USA, 2009.

59. Das, R.N.; Lin, H.T.; Lauffer, J.M.; Markovich, V.R. Printable electronics: Towards materials development and device fabrication. Circuit World 2011, 37, 38–45.

60. Ionescu, C.; Bonfert, D.; Codreanu, N.D.; Svasta, P. Investigations on current capabilities of PEDOT:PSS conductors. Proceeding of the 35th International Spring Seminar on Electronics Technology, Bad Aussee, Austria, 9–13 May 2012; pp. 59–64.

61. Gordon, P.; Torah, R; Beeby, S.; Tudor, J. The development of screen printed conductive networks on textiles for biopotential monitoring applications. Sens. Actuators A Phys. 2006, 206, 35–41.

62. Karaguzel, B.; Merritt, C.R.; Kang, T.; Wilson, J.M.; Nagle, H.T.; Grant, E.; Pourdeyhimi, B. Flexible, durable printed electrical circuits. J. Text. Inst. 2009, 100, 1–9.

63. Scalisi, R.G.; Paleari, M.; Favetto, A.; Stoppa, M.; Ariano, P.; Pandolfi, P.; Chiolerio, A. A new paradigm in surface electromyography: Inkjet printed flexible electrodes. Submitted.

64. Pacelli, M.; Loriga, G.; Taccini, N.; Paradiso, R. Sensing Fabrics for Monitoring Physiological and Biomechanical Variables: E-textile solutions. Proceedings of the 3rd IEEE-EMBS International Summer School and Symposium on Medical Devices and Biosensors MIT, Boston, MA, USA, 4–6 September 2006.

65. Huang, C.-T.; Shen, C.-L.; Tang, C.-F.; Chang, S.-H. A wearable yarn-based piezo-resistive sensor. Sens. Actuators A Phys. 2008, 141, 396–403.

66. Hertleer, C.; Grabowska, M.; Van Langenhove, L.; Catrysse, M.; Hermans, B.; Puers, R.; Kalmar, A.; Van Egmond, H.; Matthys, D. Toward a Smart Suit. Proceeding of Wearable Electronic and Smart Textiles, Leeds, UK, 11 June 2004.

67. Gough, P. Electronics and Clothes: Watt to Wear? Proceeding of Wearable Electronic and Smart Textiles, Leeds, UK, 11 June 2004.

68. Ashruf, C.M.A. Thin flexible pressure sensors. Sens. Rev. 2002, 22, 322–327.

69. Rothmaier, M.; Luong, M.P.; Clemens, F. Textile Pressure Sensor Made of Flexible Plastic Optical Fibers. Sensors 2008, 8, 4318–4329.

70. Eleksen Ltd. Available online: http://www.eleksen.com (accessed on 12 January 2014).

71. Logitech Inc. Available online: http://www.logitech.com (accessed on 12 January 2014).

72. Pressure Profile Systems, Inc. Available online: http:// www.pressureprofile.com/presStrip.php (accessed on 11 January 2014).

73. Swallow, S.; Peta Thompson, A. Available online: http:// www.intelligenttextiles.com (accessed on 1 January 2014).

74. Viry, L.; Levi, A.; Totaro, M.; Mondini, A.; Mattoli, V.; Mazzolai, B.; Beccai, L. Flexible three-axial force sensor for soft and highly sensitive artificial touch. Adv. Mater. 2014, 2659–2664.

75. Windmiller, J.R.; Wang, J. Wearable Electrochemical Sensors and Biosensors: A Review. Electroanalysis 2012, 1, 29–46.

76. Benito-Lopez, F.; Coyle, S.; Byrne, R.; Smeaton, A.; O'Connor, N.; Diamond, D. Pump Less Wearable Microfluidic Device for Real Time pH Sweat Monitoring. Proceeding of the Eurosensors XXIII Conference, Lucerne, Switzerland, 6–9 September 2009.

77. Nishide, H.; Oyaizu, K. Materials science. Toward flexible batteries. Science 2008, 319, 737–738.

78. Infineon Technologies AG. Available online: http://www. infineon.com (accessed on 2 February 2014).

79. Swallow, L.M.; Luo, J.K.; Siores, E.; Patel, I.; Dodds, D. A piezoelectric fibre composite based energy harvesting device for potential wearable applications. Smart Mater. Struct. 2008, 17.

80. Wei, Y.; Torah, R.; Yang, K. Screen printing of a capacitive cantilever-based motion sensor on fabric using a novel sacrificial layer process for smart fabric applications. Meas. Sci. Technol. 2013, 24.

81. Wei, Y.; Torah, R.; Yang, K.; Beeby, S.; Tudor, J. Screen printed capacitive free-standing cantilever beams used as a motion detector for wearable sensors. Procedia Eng. 2012, 47, 165–169.

82. Bai, S.; Zhang, L.; Xu, Q.; Zheng, Y.; Qin, Y.; Wang, Z. Two dimensional woven nanogenerator. Nano Energy 2013, 2, 1–5.

83. Lee, Y.-H.; Kim, J.-S.; Noh, J.; Lee, I.; Kim, H.J.; Choi, S.; Seo, J.; Jeon, S.; Kim, T.-S.; Lee, J.-Y.; et al. Wearable Textile Battery Rechargeable by Solar Energy. NanoLetters 2013, 13, 5753–5761.

84. Chiechi, R.C.; Havenith, R.W.A.; Hummelen, J.C.; Koster, L.J.A.; Loi, M.A. Modern plastic solar cells: Materials, mechanisms and modeling. Mater. Today 2013, 16, 281–289.

85. Schubert, M.B.; Werner, J.H. Flexible solar cells for clothing. Mater. Today 2006, 9, 42–50.

86. Consultancy Goose Design, "PDD Illum Project". Available online: http://www.gooseuk.com/ (accessed on 11 November 2013).

87. Unitika LTD. Available online: http://www.unitika.co.jp/e (accessed on 2 February 2014).

88. Meoli, D. Interactive Electronic Textiles: Technologies, Applications, Opportunities, and Market Potentia. MS.c. Thesis, North Carolina State University, Rayleigh, NC, USA, 2002.

89. Kim, H.; Kim, Y.; Yoo, H.J. A 1.12 mW continuous healthcare monitor chip integrated on a planar-fashionable circuit board. Proceeding of the IEEE International Solid-State Circuits Conference (ISSCC) Digest of Technical Papers, San Francisco, CA, USA, 3–7 February 2008; pp. 150–151.

90. Seulki, L.; Binhee, K.; Yoo, H.J. Planar Fashionable Circuit Board Technology and Its Applications. J. Semicond. Technol. Sci. 2009, 9, 174–180.

91. Giddens, H.; Paul, D.L.; Hilton, G.S.; McGeehan, J.P. Influence of body proximity on the efficiency of a wearable textile patch antenna. Proceeding of the 6th European Conference Antennas & Propagation (EuCAP), Prague, Czech, 26–30 March 2012; pp. 1353–1357.

92. Zhang, L.; Wang, Z.; Psychoudakis, D.; Volakis, J.L. Flexible Textile Antennas for Body-Worn Communication. Proceedings of IEEE International Workshop on Antenna Technology, Tucson, ZA, USA, 5–7 March 2012; pp. 205–208.

93. Grupta, B.; Sankaralingam, S.; Dhar, S. Development of Wearable and Implantable Antennas in the Last Decade: A Review. Proceedings of Mediterranean Microwave Symposium (MMS), Guzelyurt, Turkey, 25–27 August 2010; pp. 251–267.

94. Liu, N.; Lu, Y.; Qiu, S.; Li, P. Electromagnetic Properties of Electro-Textile for Wearable Antennas Applications. Front. Electr. Electron. Eng. China 2011, 6, 553–566.

95. Brebels, S.; Ryckaert, J.; Boris, C.; Donnay, S.; De Raedt, W.; Beyne, E.; Mertens, R.P. SOP Integration and Codesign of Antennas. IEEE Trans. Adv. Packag. 2004, 27, 341–351.

96. Moretti, A. Estudo do Brim Santista Visando Aplicações em Antenas têxteis. MS.c. Thesis, Universidade Estadual de Campinas, Campinas, Brazil, 2011.

97. Roh, J.-S.; Chi, Y.-S.; Lee, J.-H.; Tak, Y.; Nam, S.; Kang, T.J. Embroidered Wearable Multiresonant Folded Dipole Antenna for FM Reception. IEEE Antennas Wirel. Propag. Lett. 2010, 9, 803–806.

98. Ouyang, Y.; Chappell, W.J. High Frequency Properties of Electrotextiles for Wearable Antenna Applications. IEEE Trans. Antenna Propag. 2008, 56, 381–389.

99. Kaija, T.; Lilja, J.; Salonen, P. Exposing Textile Antennas for Hash Environment. Proceedings of 2010 Military Communications Conference, San Jose, CA, USA, 31 October–3 November 2010; pp. 737–742.

100. Hertleer, C.; Laere, A.V.; Rogier, H.; Langenhove, L.V. Influence of Relative Humidity on Textile Antenna Performance. Text. Res. J. 2009, 80, 177–183.

101. Matthews, J.C.G.; Pettitt, G. Development of Flexible, Wearable Antennas. Proceedings of EuCAP 2009: 3rd European Conference on Antennas and Propagation, Berlin, Germany, 23–27 March 2009; pp. 273–277.

102. Tronquo, A.; Rogier, H.; Hertleer, C.; Langenhove, L.V. Applying Textile Materials for the Design of Antennas for Wireless Body Area Networks. Proceedings of EuCap 2006: First European Conference on Antennas and Propagation, Nice, France, 6–10 November 2006.

103. Hertleer, C.; Rogier, H.; Member, S.; Vallozzi, L.; Langenhove, L.V. A Textile Antenna for Off-Body Communication Integrated into Protective Clothing for Firefighters. IEEE Trans. Adv. Packag. 2009, 57, 919–925.

104. Yan, S.; Soh, P.J. Wearable dual-band composite right/left-handed waveguide textile antenna for WLAN applications. Electron. Lett. 2014, 50, 424–426.

105. Patria. Available online: http://www.patria.fi/ (accessed on 7 June 2014).

106. TEXTRACE. Available online: http://www.textrace.com/ (accessed on 7 June 2014).

107. Brosteaux, D.; Gonzalez, M.; Vanfleteren, J. Design and Fabrication of Elastic Interconnections for Stretchable Electronic Circuits. IEEE Electron Device Lett. 2007, 28, 552–554.

108. Gonzalez, M.; Axisa, F.; Vanden Bulcke, M.; Brosteaux, D.; Vandevelde, B.; Vanfleteren, J. Design of metal interconnects for stretchable electronic circuits. Microelectron. Reliab. 2008, 48, 825–832.

109. Huyghe, B.; Rogier, H. Design and manufacturing of stretchable high-frequency interconnects. IEEE Trans. Adv. Packag. 2008, 31, 802–808.

110. STATEX. Available online: http://www.statex.de/index.php/en/ (accessed on 4 July 2013).

111. UGITECH S.A. Available online: "http://www.sprintmetal.com (accessed on 10 September 2013).

112. Bekaert Fibre Technologies. Available online: http:// www. bekaert.com (accessed on 25 September 2013).

113. Locher, I.; Kirstein, T.; Tröster, G. Routing methods adapted to e-textiles. Proceeding of the 37th International Symposium on Microelectronics (IMAPS), Long Beach, CA, USA, 16–18 November 2004.

114. Cottet, D.; Grzyb, J.; Kirstein, T.; Tröster, G. Electrical characterization of textile transmission lines. IEEE Trans. Adv. Packag. 2003, 26, 182–190.

115. Bhat, N.V.; Seshadri, D.T.; Radhakrishnan, S. Preparation, characterization, and performance of conductive fabrics: Cotton + PANi. Text. Res. J. 2004, 74, 155–166.

116. Tong, L.; Lijing, W.; Xungai, W.; Akif, K. Polymerising pyrrole on polyester textiles and controlling the conductivity through coating thickness. Thin Solid Films 2005, 479, 777–782.

117. Cochrane, C.; Koncar, V.; Lewandowski, M.; Dufour, C. Design and development of a flexible strain sensor for textile structures based on a conductive polymer composite. Sensors 2007, 7, 473–492.

118. Shyr, T.W.; Shie, J.W.; Jhuang, Y.E. The effect of tensile hysteresis and contact resistance on the performance of strain-resistant elastic-conductive webbing. Sensors 2001, 11, 1693–1705.

119. Mattmann, C.; Clemens, F.; Tröster, G. Sensor for Measuring Strain in Textile Wearable. Sensors 2008, 8, 3719–3732.

120. Coyle, S.; Lau, K.; Moyna, N.; Diamond, D.; Di Francesco, F.; Constanzo, D.; Salvo, P.; Trivella, M.; De Rossi, D.; Taccini, N.; et al. BIOTEX—Biosensing textiles for personalised healthcare management. IEEE Trans. Inf. Technol. BioMed. 2010, 14, 374–370.

121. Gorlatova, M.; Sarik, J.; Cong, M. Movers and Shakers: Kinetic Energy Harvesting for the Internet of Things. In Technical Report; Electrical Engineering Department, Columbia University: New York, NY, USA, 2013.

122. Seulki, L.; Binhee, K.; Taehwan, R.; Sunjoo, H.; Hoi-Jun, Y. Arm-band type textile-MP3 player with multi-layer Planar Fashionable Circuit Board (P-FCB) techniques. Proceeding of the International Symposium on Wearable Computers (ISWC), Seoul, South Korea, 10–13 October 2010.

123. Kim, Y.; Kim, H.; Yoo, J. Electrical characterization of printed circuits on the fabric. IEEE Trans. Adv. Packag. 2010, 33, 196–205.

124. Baker-Jarvis, J.; Janezic, M.D.; DeGroot, D.C. High-Frequency Dielectric Measurements. IEEE Trans. Instrum. Meas. 2010, 13, 24–31.

125. Salvado, R.; Loss, C.; Gonçalves, R.; Pinho, P. Textile materials for the design of wearable antennas: A survey. Sensors 2012, 12, 15841–15857.

126. Lilja, J.; Salonen, P.; Kaija, T.; De Maagt, P. Design and Manufacturing of Robust Textile Antennas for Harsh Environments Antennas and Propagation. IEEE Trans. Antenna Propag. 2012, 60, 4130–4140.

127. Salonen, P.; Rahmat-Samii, Y.; Hurme, H.; Kivikoski, M. Effect of Conductive Material on Wearable Antenna Performance: A Case Study of WLAN Antennas. Proceedings of IEEE Antennas and Propagation Society International Symposium, Monterey, CA, USA, 20–25 June 2004; pp. 455–458.

128. Hu, L.; Pasta, M.; La Mantia, F.; Cui, L.; Jeong, S.; Deshazer, H.D.; Choi, J.W.; Han, S.M.; Cui, Y. Stretchable, porous, and conductive energy textiles. Nano Lett. 2010, 10, 708–714.

CHAPTER 5

Embroidery and Related Manufacturing Techniques for Wearable Antennas: Challenges and Opportunities

Aris Tsolis [1,2], William G. Whittow [1,*], Antonis A. Alexandridis [2] and J. (Yiannis) C. Vardaxoglou [1]

[1] *School of Electronic, Electrical and Systems Engineering, Loughborough University, Loughborough LE11 3TU, UK*

[2] *Institute of Informatics and Telecommunications, National Centre for Scientific Research "Demokritos", Athens 15310, Greece*

ABSTRACT

This paper will review the evolution of wearable textile antennas over the last couple of decades. Particular emphasis will be given to the process of embroidery. This technique is advantageous for the following reasons: (i) bespoke or mass produced designs can be manufactured using digitized embroidery machines; (ii) glue is not required and (iii) the designs are aesthetic and are integrated into clothing rather than being attached to it. The embroidery technique will be compared to alternative manufacturing processes. The challenges facing the industrial and public acceptance of this technology will be assessed. Hence, the key opportunities will be highlighted.

KEYWORDS

wearable antennas; embroidered antennas; wearable technology; inkjet printing; textiles

1. INTRODUCTION

Wearable antennas technology has rapidly grown over the last couple of decades. It could be said and assumed that this technology is the future of smart garments and furthermore the future of our daily life. A wearable antenna is meant by definition to be part of the garments worn by humans or animals. Therefore, the antenna should be flexible or miniaturized so as to not inhibit the movement of the human body. Consumers will demand smaller and more intelligent communications systems which will enhance their quality of life. Smart garments due to the evolution of wearable antennas technology may find their place in our everyday lifestyle. Smart clothes will emerge in various applications including: sports, emergency workers, military, medical and space applications or even in casual daily clothes or in fashion [1].

This paper will be structured as follows. Section 2 will review the dielectric properties of textile substrates. Section 3 will compare different manufacturing techniques with particular emphasis on embroidery. Section 4 will consider particular examples and applications of wearable rigid miniaturized and textile antennas. Section 5 will address the challenges both from a technical point of view and also from the market acceptance point of view. Section 6 will consider some of the opportunities that exist in this area. An extensive (but not complete) list of references will be given at the end of this paper.

2. SUBSTRATE MATERIALS FOR WEARABLE ANTENNAS

The substrate material selection for a wearable antenna design is of crucial importance. The substrate selection requires a low loss material so as to have better chances of increased antenna efficiency when placed on the body. The work in [2] where a Planar Inverted F-Antenna (PIFA) was fabricated out of a foam sheet spacer as the substrate and copper plated nylon for conductive sections states that a low loss conductive fabric is critical to the performance of a textile antenna. For wearable textile antennas the substrate material is required to be flexible. Textile substrate dielectric properties vary with the choice of material and the frequency. The accurate characterization of the properties of a textile substrate is fundamental before designing a textile antenna. A summary of the properties that various researchers use for antenna design are outlined in Table 1.

Table 1. Substrates' dielectric properties.

Material	Permittivity (ε_r)	Loss Tangent (tanδ)	Frequency (GHz)	Reference
Denim	1.40		2.4 & 5.2	[3]
Denim	1.40		0.9 & 1.8	[4]
Leather	2.95	0.16	0.9 & 1.8	
Denim	1.70		3–12	[5]
Denim	1.80	0.07	2.4 & 5	[6]
Velcro	1.34	0.006	2.4 & 5	
Denim	1.8–2.0	0.014	14–40	[7]
Denim (black)	1.8	0.07	3.3 & 5	[8]
Velcro	1.37		2.4 & 5	[9]
Felt	1.38	0.023	2.6–3.95	[10]
Fleece	1.17	0.0035		
Moleskin	1.45	0.05		
Panama	2.12	0.018		
Silk	1.75	0.012		
Tween	1.69	0.0084		
Perspex	2.57	0.008		
PTFE	2.05	0.0017		

Table 1. (Continued)

Material	Permittivity (ε_r)	Loss Tangent (tanδ)	Frequency (GHz)	Reference
Polystyrene foam	1.02	0.00009	2.4	[11]
Felt	1.36	0.016		
Fleece	1.2	0.004		
Neoprene rubber	5.2	0.025		
Silk	1.2	0.054		
Cotton	1.54	0.058		
Leather—different types	1.8–2.4	0.049–0.071		

From the above table it is clear that textiles typically have a low relative permittivity. The work in [12] studied the effect of different textile materials on wearable antenna performance. A patch antenna for Global Positioning System (GPS) operation was designed and five antennas were fabricated [12]; all of them used copper tape for the conductive sections. The different textile materials used for the substrate were: fleece (thickness, t = 4 mm), upholstery fabric (t = 1.1 mm), vellux (t = 5 mm), synthetic felt (t = 4 mm) and cordura (t = 0.5 mm). Non-conductive textile materials have low relative permittivity which reduces the surface waves (for patch antennas) and improves antenna impedance bandwidth. The measured dielectric constant of these textile materials at 1.575 GHz was between 1.1 and 1.7. Therefore, fabric substrate antennas are somewhat larger in physical dimensions than rigid antennas due to the low value of permittivity. Cordura is made by Delcotex, Germany. The name of the fabric is Delinova 200, which is woven from Cordura fibers. Cordura is a polyamide fiber with a nylon and Gore-Tex membrane. Delinova 200 is a textile impregnated with fluorocarbon and coated with polyurethane. Delinova 200—Cordura fabric proved to be an interesting fabric for textile antennas because of its strength, constant thickness and high water resistance properties [1,12].

The effect of using different textile materials as the substrate of a patch antenna was studied in [13]. Four antennas were designed and fabricated. The different textiles that were used for antenna fabrication are: wash cotton, curtain cotton, polyester and polycot. The polyester patch antenna yielded the best antenna performance in terms of gain and efficiency (polyester patch antenna yielded measured gain equal to 9.6 dBi compared to 7.2–7.5 dBi with the other antennas) [13]. The polyester had the lowest loss tangent compared to the other three textiles that were used for antenna fabrication. Additionally, as shown in [14,15] fleece fabric provides sufficient thickness for an adequate bandwidth. The low permittivity of fleece allows design of textile wearable antennas with large gain and high efficiency which claims fleece fabric as a very good candidate for textile substrate material.

The firmness of a textile material substrate can affect the performance of the textile antenna. An embroidered textile antenna (Sierpinski Carpet Antenna) was designed in [16] using fractal antenna technology. Two antennas were fabricated. One antenna was embroidered on denim and the other on felt. The first one showed the best performance—denim fabric has a firm surface compared to felt fabric that has a soft fluffy surface which gives disadvantages to felt—fluffy surfaces are easily compressed and the variable thickness affects the antenna properties. Both fabricated antennas are very good candidates for wearable applications. For good antenna performance homogeneity of the textiles is a requirement [10]. The accurate characterization of the textile properties is also a requirement for good agreement between the antenna design and manufacture [10]. In [17] a novel method to measure the relative permittivity of textile substrates and eventually to design and manufacture a textile patch antenna is described and proposed. Once a patch has been designed by using an estimated value for the dielectric constant and then when the antenna is manufactured, a different resonant frequency occurs which means that the initial estimation was wrong. Then there is a need of tuning the permittivity and using the right value for the design. One method of characterizing the dielectric properties is by using split post dielectric resonators at the respective frequencies as seen in Figure 1. The dielectric properties can be measured for example using a split post dielectric resonator [18]. However, in practice an effective permittivity can be found by comparison with measurements.

(a) (b)

Figure 1. (**a**) Q-meter measurement setup measuring a piece of felt by using the 1.9 GHz split post resonator; (**b**) Resonator-Vector Network Analyzer (VNA) setup measuring a piece of felt by using the 1.1 GHz split post resonator.

A novel polymer substrate was developed, produced and used as an RF substrate [19]; it is known as Polydimethylxiloxane (PDMS) ($\varepsilon_r = 3$, tan $\delta < 0.02$ from 0–1 GHz). PDMS has a good mechanical flexibility and low loss performance. By adding ceramic powder to the PDMS, the polymer composite is capable of providing tunable permittivity.

3. MANUFACTURING TECHNIQUES FOR WEARABLE ANTENNAS

The design and moreover the manufacture process of a wearable antenna is of crucial importance in terms of antenna performance and manufacturing time. The manufacture process of wearable antennas should guarantee good agreement with the design and simulation results and should lead to robustness and repeatability of the wearable antenna. This arises from the fact that the wearable antenna is supposed to be embedded on the garments and it will operate under different conditions (movement of the wearer, weather conditions, temperature, bending and crumpling conditions). In addition, washability is a requirement in the case of embroidered textile antennas where the goal is that these antennas should be part of the clothes. Embroidered antennas are a great opportunity to connect wearable antenna technology with the industry of textiles. In this section, the manufacturing techniques used so far on wearable antennas are described and there are two categories: (1) rigid and (2) flexible (often textile) antennas.

3.1. Rigid Wearable Antennas

The manufacture process of rigid wearable antennas follows the conventional techniques of printing or constructing antennas for example by using etching. Rigid antennas worn on the body can be miniaturized so as to minimize the inconvenience to the user as conventional antennas are not practical, see Figure 2. These rigid antennas sometimes need to be manufactured in a curved contoured shape so as to meet the wearable requirement [20]. Another difficulty is the small size of these antennas and their complex structure [21].

3.2. Flexible Textile Wearable Antennas

Textile antenna manufacturing techniques can be divided into the following categories: (**1**) Thin and uniform metallization layers (i.e., copper or silver tape, foil) attached to the non-conductive textile fabric [22]; (**2**) the use of conductive textile yarns to weave or knit the conductive patterns of the antenna and then attach or stitch them onto the non-conductive textile substrate [22]; (**3**) the use of conductive textile yarns to embroider the conductive patterns of the antenna on the non-conductive textile substrate [22]; (**4**) Inkjet and screen printed printing on non-conductive textile materials.

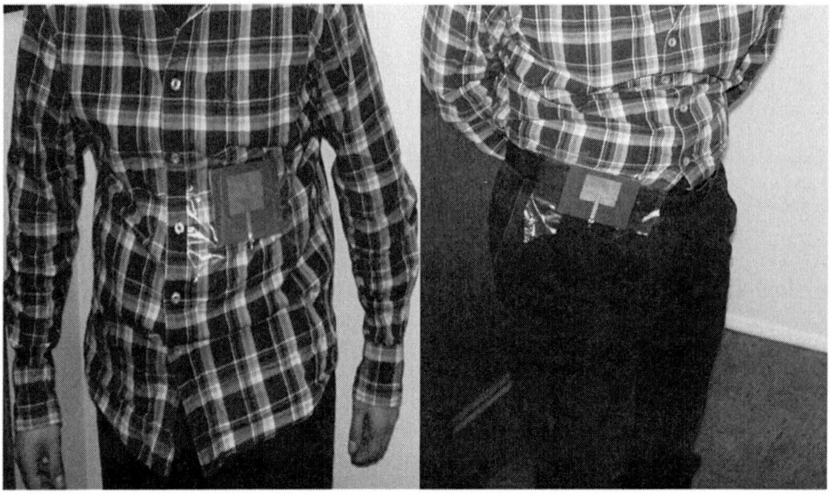

Figure 2. (**a**) 2.5 GHz patch antenna positioned at the left side of the thorax under the chest; **(b)** 2.5 GHz patch antenna positioned on the trunk.

Before moving into the details of each technique it is important to note that different conductive textile materials (i.e., copper tape, knitted copper, etc.) can result in different antenna performance. The effect of different conductive materials on wearable antenna performance is presented in [23]. Six different antennas were fabricated. For all the antennas, the substrate that was used was fleece fabric (thickness = 3.5mm, ε_r = 1.1). The different conductive sections of the antennas were: solid copper tape; knitted copper tape; vertically cut pieces of copper tape; horizontally cut pieces of copper tape; horizontally cut and soldered copper tape; aracon fabric. Conductive tapes and fabrics are both acceptable solutions for the conductive sections [23]. Though, not all copper tape and fabric configurations are appropriate for textile antennas. The conductive fabric is required to have a good conductivity and needs to have a high density of conducting fibers. A principle derived from this study [23] is that there can be discontinuities in the conducting shape from cutting the copper tape as long as these discontinuities are parallel to the surface current (vertically cut copper tape). When discontinuities are perpendicular to the current flow direction there will be reflections of the electromagnetic fields at the interfaces [23]. Additionally the use of different conductive materials (conducting nylon, embroidered conducting thread, conducting paint) as conductive sections and different constructing methods of a textile spiral antenna were presented in [24].

3.2.1. Thin and Uniform Metallization Conductive Sections

This technique is the fastest when it comes to the manufacture of one or a small number of antennas but when it comes to the mass production of textile antennas it is probably a slower than other techniques due to the manual work which is required. Generally, this technique is reasonable for experimental antenna prototyping but may not be ideal for longer term solutions. It is very easy to attach copper tape or foil on a textile substrate but it is also possible that the copper tape might be detached while bending or even due to environmental conditions (i.e., humidity, heat) (Figure 3).

One solution to avoid having the textile antenna permanently integrated on the garment is to use Velcro connectors [25]. In [25] a wideband wearable textile antenna is presented where the fabricated antenna is integrated on a jacket at the inside side of it by using Velcro. This means that the antenna can be easily physically attached and detached to and from the jacket by using Velcro strips.

3.2.2. Woven or Knitted Conductive Sheets

This technique requires more time to construct a single antenna but may allow an element of automation for mass production. This can result in a more tolerant and flexible antenna structure compared to using copper tape. We will include woven metallic cloths, in this section. An example of this is Nora Dell which is a textile coated with nickel and silver to from a highly conducting flexible sheet (Figure 4).

Figure 3. Textile patch antenna (substrate: felt, conducting sections: copper tape) with partially detached copper tape due to environmental conditions.

Many papers considering textile antennas using woven or knitted conductive sheets have been published and will be reviewed in this section. The first compact fabric antenna design for commercial smart clothing was presented in [26]. This compact antenna is a patch antenna using fleece fabric as substrate and knitted copper for the conductive sections. In [27] a nickel plated woven textile, which has a high tolerance against oxidation and corrosion, is stable and has a very good conductivity, was attached on a spacer by using ammonia-based textile glue. The textile glue showed negligible influence on the sheet resistance since the plated textile is densely woven [27]. In [15] two antennas manufactured by using two different methods of attaching conductive Flectron onto Fleece fabric were considered. The layers in the first antenna were attached to each other by using a glue stick and the layers have been additionally stitched together using non-conducting threads to enhance the robustness of the antenna. For the second prototype an adhesive sheet which melts when ironing was used to attach the layers together. The first antenna showed better performance in terms of return loss and bandwidth. It can be assumed that the adhesive sheet melting inserts more losses into the antenna compared to the glue stick.

Figure 4. Nora Dell conductive cloth.

In [28] four different methods were used to attach the conductive woven textile on the non-conductive textile were examined: (i) liquid textile adhesive (golden fix); (ii) point wise application of conductive adhesives; (iii) sewing and (iv) adhesive sheets which melt when ironed. The 4th attachment method yielded the best results out of the four methods; it barely affected the substrate (felt) thickness which helped the antenna to maintain the geometrical designed dimensions [28], also described the design of woven and knitted conductive textiles, the material selection criteria and the characterization of the material properties (sheet resistance of the conductive textiles and the permittivity, tangent loss of the non-conductive textile). The effect of various weave patterns of conductive cloths in combination with the non-conductive dielectric substrate is presented in [29,30]. This was studied because not all fabric patterns are symmetric. There are some fabrics with asymmetrical patterns on each side. This means that some conductive fabric patterns have the majority of the conductive threads on one side of the fabric. In these works ([29,30]), the cases where the side of the fabric (conductive) with the majority of the conductive threads was directly attached onto the dielectric non-conductive substrate yielded less dielectric loss and better conductivity compared to the cases where the side of the fabric (conductive) with the minority of the conductive threads was directly attached on the dielectric non-conductive substrate.

In [31] an aperture coupled patch antenna using two different textiles (fleece and felt) for the substrates was proposed, designed and manufactured. Additionally, two different conductive textiles were used for the patch, ground and the microstrip line. For the patch, Shieldit was selected because it has an additional adhesive layer which made it easier to attach the patch to the substrate. For the ground and the microstrip line Flectron was used because of its smoothness and low surface resistivity. The two layer antenna allowed an independent selection of the antenna substrate. Fleece fabric was selected for the antenna substrate because it has permittivity close to 1 and a low loss tangent which is optimal for the radiation efficiency. Felt was selected for the microstrip feed line substrate. Felt has a higher permittivity and by selecting a thinner material the radiation losses are restricted.

A new design of a woven textile antenna covered with a protective cloth (e.g., waterproof, insulation) was presented in [32]. Finally a novel grid microstrip woven antenna design and fabrication method is presented in [33]. The substrate of the antenna was a 3D woven glass fiber fabric and the conductive yarn that used was copper yarn. The conductive yarns were orthogonally woven to produce a grid structure.

3.2.3. Embroidery

The traditional embroidery process creates aesthetic shapes using colored threads on a base textile material. By using specialist conducting threads, antennas can be embroidered onto the base textile. Although the basic principles are the same, the technology has advanced to enable a digital image to be directly embroidered using a computer aided embroidery machine (Figure 5). The conducting threads must exhibit suitable flexibility and strength so as not to be broken by the high tensions required in the embroidery machine.

Of great importance before designing and embroidering the antenna is knowledge of the properties (i.e., DC resistance, conductivity) of the conductive yarns to be used [34]. Once the conductive yarns are characterized then it is easier to find techniques to improve the performance of the antenna; for example using a higher density of stitching. Conductive threads with high resistance yield poor S21 results in transmission line measurements [34]. After the characterization of the conductive yarns and of the textile substrate, the design of the embroidered textile antenna (Figure 6) comes next. The design and simulation method to model wearable antennas composed of embroidered conductive threads is presented in [35]. A detailed simulation method is required to model the directional and discrete nature of the threads [35] (Figure 7).

Figure 5. Computer aided embroidery machine at Loughborough University.

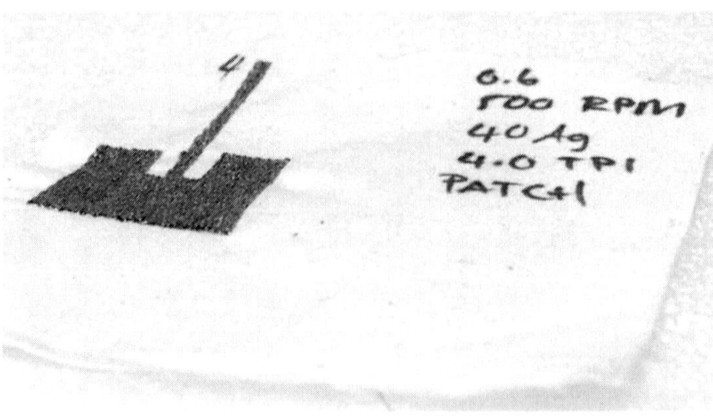

Figure 6. Embroidered patch antenna.

Figure 7. Embroidered patch antenna magnified to observe the air gap between the conductive threads.

The challenges of fabricating embroidered antennas are outlined in [36]. Such challenges include: selecting the most suitable conductive thread in terms of conductivity; strength; flexibility and the assessment of the behavior of the conductive threads when they are stitched to form an approximately continuous object so as to improve the efficiency of the patch antenna [36]. The effect of the stitch direction and stitch density on the performance of the antenna has been extensively studied and described in [37]. The understanding of the current flow on a designed patch antenna is a fundamental requirement to choose the stitch direction (Figure 8). Higher antenna efficiencies can be obtained by ensuring the principle current flow is parallel to the stitch direction. This can lead to further challenges when the design is required to also work at higher modes where the current is in the perpendicular direction and for more complicated designs where the current flows in different directions.

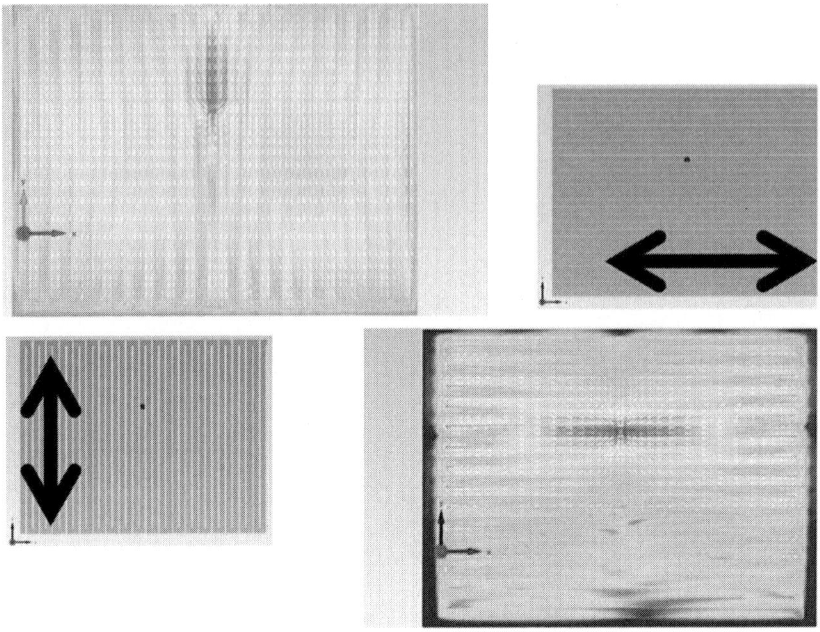

Figure 8. Simulated surface current of two different stitch direction patch antennas.

The effect on antenna performance caused from different stitching geometries is studied and presented in [38]. The effect of the stitch spacing on the gain and directivity was examined. Generally, the closer the stitching spacing, the higher the antenna efficiency will be. However, this comes at the expense of reduced flexibility and increased length of thread which directly leads to higher manufacturing costs.

The effect of the type of stitching on the dipole antenna performance is described and demonstrated in [39]; where dipole tag-antennas embroidered with different thread densities and two different stitch patterns were considered. The embroidering aspects of Ultra High Frequency (UHF) Radio-Frequency Identification (RFID) antenna were presented in [40]. Professor Volakis and his colleagues have contributed to the field of embroidered antennas and RF electronics

[41,42,43,44,45,46,47,48]. Throughout these publications novel conformal embroidered antennas and novel embroidered sensors for medical applications are presented. Novel embroidered microstrip lines were developed and the fabrication procedure of e-fiber transmission lines was proposed. An ultra-wideband embroidered textile antenna is presented in [49]. Due to the usage of conducting thread, the antenna is flexible; can facilitate washing and can be an attractive solution for a wearable device.

3.2.4. Inkjet and Screen Printed Antennas

Inkjet and screen printing [50] can also be used to create conducting sections of wearable antennas. However, the substrates are generally paper or Kapton while textiles are not ideal printing substrates. As the name suggests screen printing requires a mask to be made for each design and therefore is less practical for different individual designs [51,52]. RF transmission lines have been previously screen printed onto a cotton substrate [53].

Inkjet printing does not require a mask and designs can be created within minutes of receiving the computer file containing the geometry and hence enables great manufacturing flexibility. Typically, silver nanoparticles in solution are used to make thin conducting lines which are approximately 1 μm thick. Therefore, printing on rough surfaces such as textiles is very challenging. In addition textiles are porous materials which make it more challenging to create continuous conducting lines [54]. Other challenges of inkjet printing on textiles include achieving a continuous conducting track with a high conductivity; robustness to stretching and inherent movement and resilience to high temperatures required to remove the non-conducting solvent from the ink.

These challenges can be overcome by first screen printing an interface layer onto fabrics. This process is outlined in [55]. The inherent surface roughness of the cotton fabric is reduced by using screen printed interface layer and enables the printing of antennas with reasonable efficiencies with only one or two layers of ink. This process has been applied to dipoles [55] and patch antennas [56].

Antenna performance can be improved by printing multiple layers but this increases material costs and manufacturing time and reduces the resolution of the lines. As the ink layer is thin, skin depth issues must be considered. This is likely to limit the applicability of inkjet printing on fabrics to higher frequencies.

3.3. Comparison of Embroidery with Other Techniques

Embroidery is advantageous compared to the others because embroidery machines already exist in industry; so it is easier to apply this technique for mass production of garments with integrated embroidered textile antennas. As the currents in embroidered antennas prefer to flow along threads rather than from thread to thread, embroidery may naturally lend itself to linear antennas such as dipoles or spirals. These types of structures can be very hard to fabricate using copper tape or Nora Dell cloths. The design of spirals or dipoles as opposed to patch antennas will also reduce the length of thread required and hence the cost of the antennas. Embroidery uses specialized thread which contains silver which is expensive. Note, the same is true for inkjet printing and Nora Dell. Therefore, the level of wastage of using manufacturing techniques will strongly affect the manufacturing costs.

Embroidery allows repeatable geometries to be made via the computerized embroidery machines. As no mask is required, embroidery can also make one-off bespoke items. Repeatable structures as a Frequency Selective Surface (FSS) structure (Figure 9) can be embroidered by copying the original shape [57]. Fractal antenna technology [16] to design wearable textile antennas could result in attractive and compact designs, however, the accuracy of embroidery is limited to approximately one millimeter. This can be partially overcome by the use of computer aided embroidery. Additionally with embroidery the use of glue is not always a requirement to attach the textile layers together [58]. This can enhance the washability of the garment with the integrated antenna.

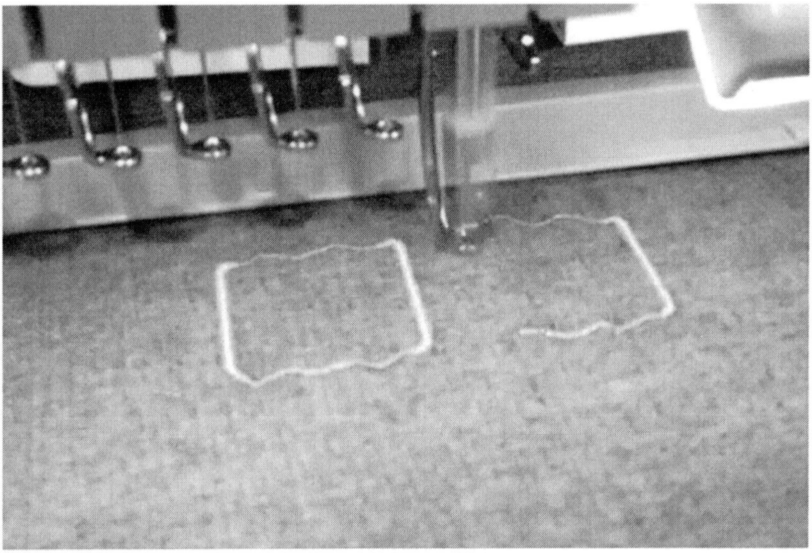

Figure 9. Embroidery machine creating FSS structure.

4. SPECIFIC EXAMPLES AND APPLICATIONS OF WEARABLE ANTENNA DESIGNS

A large number of wearable antennas have already been proposed for many different applications including miniaturized rigid and flexible textile antennas. Examples of rigid miniaturized antennas are outlined in the first half of this section. A shorted spiral-like patch antenna operating at 430 MHz suitable for military Radio Frequency (RF) Communications was presented in [21]. This antenna was etched on FR4. The use of an electromagnetic bandgap (EBG) structure makes this antenna suitable for wearable applications by reducing the power radiated towards the wearer. The overall size of this antenna is 11.4 cm × 7.6 cm, which is small compared to the wavelength at the resonant frequency. In [20] a dipole and a spiral antenna for military wearable applications (100–500

MHz) were proposed. These antennas are unobtrusive and low profile. In [59] three antennas: (a) $\lambda_g/4$ antenna with ground shield; (b) a dipole V antenna and (c) a square dipole antenna were made out of copper with a glass—epoxy substrate (ε_r = 4.8) and proposed for operation 868 MHz, for medical tele-monitoring applications. These antennas would be more suitable for a patient to wear if they were textile which will make them more flexible and comfortable. In [60] Salonen presented a dual band (900 MHz & 2.4 GHz) Planar Inverted F-Antenna (PIFA) antenna for wearable applications suitable for the Global System for Mobile Communications (GSM) and for short range Bluetooth bands. The antenna was designed as a PIFA so as to radiate away from the human body. A dual mode antenna for on/off-body communications (10 MHz/2.45 GHz) was proposed in [61]. This antenna consists of an L-shaped slit loaded for the 2.45 GHz band (off-body link) connected with an electrode which is mounted on the body and is suitable for the 10 MHz (on-body) link. In [62] a four arm spiral slot patch antenna for radio telemetry capsules was proposed for operation at 915 MHz. In terms of miniaturization a reduced ground plane shorted patch antenna for on-body communications at 2.45 GHz was proposed by Scanlon et al. [63] with a miniaturized (~λ/5) ground plane. A monopole antenna mounted vertically on the body radiates the same way (parallel radiation towards body surface) as this shorted patch [63]. However, this shorted patch proved to be more suitable for the on-body link because it is compact. In [64] a miniaturized diversity antenna dedicated to wireless body area network was proposed. The antenna was printed on FR4. The combination of a PIFA and a top-loaded monopole yielded distinct patterns. The strong isolation observed between the broadside and the end-fire radiation of the antenna is a significant feature which limits the correlation between the received signals. Other miniaturized wearable antenna designs are detailed in [65,66,67].

Smart clothes evolution will eventually result in the establishment of textile antennas in widespread use. This may mean that clothes and wearable textile antennas or electronics will be designed as one piece. Examples of wearable textile antennas are outlined in the second half of this section. A textile antenna suitable for fire fighters jackets was presented in [68]. The substrate shell of jacket is made of aramid fabric which is fireproof. In [69,70] a flexible dual band PIFA was proposed for commercial smart clothing. An inexpensive textile patch antenna for off-body wireless sensor communications for monitoring patients at the 915 MHz Industrial, Scientific and Medical (ISM) radio band is presented in [71]. Zelt was used for the conducting parts and felt for the substrate of this antenna. The cost per square meter of Zelt and Felt is equal to $20 [71]. With 1 square meter of these two materials we can fabricate about 5–6 such antennas. A dual band E-shaped patch textile antenna made from felt and copper tape is presented by Salonen in [72]. This antenna operates at 1850 MHz (BW = 100 MHz) and 2450 MHz (BW = 110 MHz). The wearer with this antenna can access GSM1900 and Wi-Fi communication links. Additionally a dual-band wearable textile antenna for the same communication links was proposed in [73]. A novel circularly polarized textile antenna for personal satellite communications is proposed in [74]. Two textile antennas using Nora Dell for conducting sections and Nomex for the substrate are designed in [75]. The two antennas are proposed to be integrated in an extravehicular suit (astronaut suit) for space network applications. A complementary-8 shape e-textile antenna element can be used as part of a body worn communication and navigation antenna system that supports many different bands, including those used by both 802.11 and 802.16 bands. A system of six complementary-8 shape e-textile antennas shifted by 90 degrees can be used for polarization diversity and omnidirectional coverage most of the observable sphere surrounding the astronaut. This is an important requirement for the astronaut communication and navigation. In [76] a fabric equiangular spiral antenna using Nora Dell for the conductive sections is fabricated and presented. The earliest wearable active textile receiving antenna in the 2.45 GHz ISM band was proposed in [77,78]. A low noise amplifier (LNA) is fabricated on a hybrid textile fabric and positioned directly underneath a wearable patch antenna. In [79] a dual band coplanar waveguide feed patch antenna was designed. Zelt is used for the conductive parts of the antenna and felt for the substrate. The use of an EBG substrate under the patch antenna improves the gain and reduces the back radiation by at least 13dB making the antenna suitable for wearable applications and more tolerant to the effects of the lossy human body in terms of antenna efficiency. Additionally, the improvement on antenna performance by the use of an EBG is presented in [80] where the effect of bending the textile patch antenna is fully examined. It is hypothesized that the EBG structure which inevitably adds an extra height to the patch antenna could be hidden into clothing. Another triangular textile patch antenna over an EBG structure is presented in [81]. In [82] the on-body improvement performance of a dual-band wearable textile by using an EBG structure was presented. Two novel Ultra Wide Band (UWB) textile antenna designs were proposed in [83]. The conducting parts of the antennas were made with Shieldit and the substrate is a thick felt. These antennas are fed by a coplanar waveguide. Because of the lack of a ground plane between the antenna and the human body the placement of the antennas on the body degraded the S11 at most locations. The dielectric coupling between the antennas and the body introduced a down shift of the resonant frequencies. The coupling with the lossy body is expected to degrade the efficiency and the gain of these antennas. Therefore, the use of an EBG structure as previously referred will improve the performance of the coplanar waveguide feed textile antennas. The process of designing a printed textile monopole antenna is presented in [84] which is lightweight and has a very simple structure to fabricate. One disadvantage of a printed monopole is that due to its short partial ground plane its performance is highly affected and degraded by the presence of the human body. This last point renders the printed textile monopole a moderate candidate for wearable applications. A broadband waveguide slot antenna is proposed where denim was used as the substrate filling the waveguide—this proposed waveguide slot antenna is cheap to manufacture [8]. A novel wearable substrate integrated waveguide (SIW) antenna fabricated entirely from textile materials is presented in [85]. The SIW-on-textile integration results in an antenna exhibiting high robustness against bending, low influence of the human body and high front-to-back ratio. In addition this antenna is light weight, flexible and low cost, thus making this antenna well suited for on-body use. Additionally a new wearable textile antenna based on a half-mode substrate integrated cavity was proposed in [86].

Various embroidered textile antennas have been proposed for different applications. A low frequency (COSPAS/SARSAT satellite beacon—406 MHz) embroidered spiral antenna is proposed in [87]. An embroidered wearable multi-resonant folded dipole antenna for FM reception was fabricated in [88]. Also a UWB embroidered antenna design where stainless steel thread was used for the conductive thread is presented in [89]. GSM and Wi-Fi embroidered textile antennas were fabricated onto regular fabrics using an automated embroidery procedure with high density stitching [90]. The proposed antennas offer user-comfort and flexibility.

A distributed body-worn transceiver system for mobile communications with the use of electro-textile antennas was proposed in [91]. Two major features causing the popularity increase for the distributed body-worn system are: (1) the high efficiency of the electro-textile antennas that can be seamlessly embedded into human clothing and (2) significant system diversity gain due to spacing among antennas distributed on the body.

5. CHALLENGES OF WEARABLE TEXTILE ANTENNAS

Once the wearable textile antenna is manufactured then the step of characterizing it is the next in line. Characterize how the wearable textile antenna operates in Free Space and when mounted on a human body. This section addresses some of the challenges of measuring wearable antennas. Initially it is useful to measure the antenna performance (gain, impedance matching (S11), directivity, efficiency, etc.) in free space [92,93,94]. Then the wearable textile antenna will generally operate under the movement of human body, some specific deformations on the antenna such as bending (on cylinder or on a human arm) [13,95,96,97,98] and crumpling [99,100] must be considered. The antenna should be measured under different environmental conditions such as high humidity to represent perspiration or exposure to water [101,102]. Additionally, the embroidered textile antennas are supposed to be part of the garment. So testing the durability of the antenna in terms of performance after washing is required [103]. Also the repeatability in terms of performance of a specific textile antenna should always be examined [104].

Measurements on a real human body [105,106] or with a human phantom [107,108,109] should take place. The antenna on-body or on-phantom should be measured in terms of S11, near-field, far-field [110,111,112]. The antenna-body interactions need to be fully characterized. Measurements can take place in an anechoic chamber or in a real environment (outdoors [113] and indoors). Measurements are the final and the most important step to characterize an antenna system. All practical applications rely on measurements results rather than purely simulations.

Measurements of wearable antennas are particularly challenging. The conductivity of the flexible materials is often lower than copper. Dielectric properties of the substrate are generally not well known and can vary from different suppliers or even different batches of the same supplier. As the permittivity is related to the density of fibers it can even change within the same substrate. The thickness of the substrate may changes depending on the method of attaching the different elements which is often an arbitrary non-repeatable process such as adding the same amount of glue across the substrate. The point where the connector meets the antennas is critical in all antenna designs. However, in wearable antennas, this sensitive point depends on the position and orientation of the antenna. These parameters increase the difficulty of wearable textile antenna measurements and make it a real research challenge.

Additionally, when antennas are placed in close proximity to the human body, the specific absorption rate (SAR) must comply with International Commission on Non-Ionizing Radiation Protection (ICNIRP) and IEEE limits [114]. This means that the body can only be exposed to limited levels of electromagnetic fields. There is still public concern about possible health issues and EM radiation. Hence, while the majority of people are happy using mobile phones a new type of antenna embedded in clothing is likely to cause some level of public debate. Wearable antennas are clearly close to the body and the SAR must be considered via simulations and measurements. The emitted power of a device is related to the communication distance. Therefore, Bluetooth wearable devices that connect to a mobile phone will use low power levels and hence have very low SAR values. Whereas devices communicating with mobile phone base stations or even satellites will transmit more power and hence have higher SAR values.

The connection between the wearable antenna and the electronics is arguably the greater weakness in wearable electronics. The majorities of published papers do not consider this and use an SMA connector and a coaxial cable. Having a rigid connector negates some of the advantages of having a textile antenna. Recently, RF measurements have demonstrated the feasibility of using Hook and Loop as an RF connection mechanism at the low GHz frequency range [115]. This has the advantage that the antenna can be disconnected from the electronics and/or clothing to allow washing or for the electronics to be placed in a different item of clothing. However, considerable work is needed to find the optimal solution to this challenge.

Finally, manufacturing costs are still a challenge for wearable antennas. The specialized threads for embroidery typically cost \$1.50 per meter and although prices will decrease, the silver content means that costs may not decrease until other alternative materials can be found. As a rough guidelines embroidered antennas operating at 2.4 GHz (~6 cm × 4.5 cm) can contain 5 m of thread. The cost of Nora Dell and inkjet inks are also dependent on the cost of the silver raw materials.

6. FUTURE OPPORTUNITIES

Futurists and antenna designers have been discussing the idea of wearable technology for nearly two decades. Generally, the integration of wearable technology into main stream society has been slower than expected. This was due to a combination of the technology not being ready and insufficient market pull.

Current technology is driven by the need for data and connectivity. There were one billion smart phones sold in 2013 [116]. This is evidence that people not only want technology, but they will spend money if it enhances their quality of life. Sensors that communicate voice, video, physiological environmental and location data will become increasingly ubiquitous. It is predicted that there will be 50 billion devices connected to the internet by 2020 [117].

The feeling that wearable technology is on the cusp of going mainstream was exemplified by the level of interest at the Consumer Electronics Show in Las Vegas in 2013 where wearable technology was one of the main talking points [118]. There are now many devices on the market integrated into wrist bands that monitor the calories a person burns per day. The market for wearable technology is estimated to be worth more than \$5 Billion by 2016 [119] and this market includes medical, fashion, sports, military and emergency services [32]. Other estimates have this value as high as \$19B by 2018 [120]. 400 million smart watches are expected to be sold by 2020 [121]. Intelligent glasses with a small screen in front of the eye [122] are now available. Major industrial companies are moving into the wearable technology area [123].

Currently, two thirds of hospital beds in the UK are occupied by people over 65. Furthermore, it is predicted that 10 million people alive in the UK today will live to the age of one hundred [124]. These numbers indicate that the current medical care of the ageing population is not financially sustainable without technological assistance. Telemedicine and remote monitoring can allow patients to be monitored in their own homes or with less medical supervision which will give more comfort to the patients. An example of this is electrocardiography (ECG) systems which monitor heart rate of patients by using antennas integrated into plasters [125]. There are also big opportunities for both recreational and professional athletes.

The requirement for wireless connectivity means that every device may have at least one antenna. Currently the electronics and battery are rigid items and therefore a rigid antenna can also be included without inconvenience, especially at 2.4 GHz where PIFAs are small. Many observers believe that the next step will be for this technology to become truly wearable and be integrated directly into clothing where textile antennas will be required. Electronics is continuing to shrink in size and emerging technologies may allow the textiles themselves to harvest energy [126]. RFID tags in clothing are a current suitable candidate for fabric antennas as no battery is required and the RFID chip is very small so the use of flexible antennas will improve user comfort. Ongoing research into flexible batteries and flexible electronics will further drive the desirability for flexible antennas.

Textile wearable antennas have the advantage that they can be integrated directly into clothing; are aesthetic and do not need to be handheld which allows the technology to exist without the user consciously being involved. By making antennas out of textiles, the size limitations become the torso size and not just the size of the device. This relaxation of the size constraints can increase the gain and bandwidth. This has particular relevance to lower frequencies where the wavelengths are larger.

AUTHOR CONTRIBUTIONS

Aris Tsolis wrote the majority of this paper. Structural and technical advice as well as editing was provided by William G. Whittow, Antonis A. Alexandridis and J. (Yiannis) C. Vardaxoglou.

REFERENCES

1. Hall, P.; Hao, Y. Antennas and Propagation for Body-Centric Wireless Communications, 1st ed.; Artech House: London, UK; Boston, MA, USA, 2006.
2. Massey, P.J. Mobile Phone Fabric Antennas Integrated Within Clothing. In Proceedings of the 11th International Conference on Antennas and Propagation, Manchester, UK, 17–20 April 2001; pp. 344–347.
3. Sanz-Izquierdo, B.; Batchelor, J.C. A Dual Band Belt Antenna. In Proceedings of the International Workshop on Antenna Technology: Small Antennas and Novel Metamaterials (iWAT 2008), Chiba, Japan, 4–6 March 2008; pp. 374–377.
4. Gaspar, D.; Moreira, A.A. Belt Antenna for Wearable Applications. In Proceedings of the IEEE Antennas and Propagation Society International Symposium (APSURSI '09), Charleston, SC, USA, 1–5 June 2009; pp. 1–4.
5. Sanz-Izquierdo, B.; Batchelor, J.C.; Sobhy, M.I. Compact UWB Wearable Antenna. In Proceedings of the Loughborough Antennas & Propagation Conference (LAPC 2007), Loughborough, UK, 2–3 April 2007; pp. 121–124.
6. Sanz-Izquierdo, B.; Batchelor, J.C.; Sobhy, M.I. Button antenna on textiles for wireless local area network on body applications. IET Microw. Antennas Propag. **2010**, 4, 1980–1987.

7. Rezgui, N.; Harmer, S.W.; Luklinska, Z.; Bowring, N.; Ren, G. Determination of the complex permittivity of textiles and leather in the 14–40 GHz millimeter-wave band using a free-wave transmittance only method. IET Microw. Antennas Propag. **2008**, 2, 606–614.

8. Sanz-Izquierdo, B.; Wu, L.; Batchelor, J.; Young, P. Textile Integrated Waveguide Slot Antenna. In Proceedings of the IEEE Antennas and Propagation Society International Symposium (APSURSI), Toronto, ON, USA, 11–17 July 2010; pp. 1–4.

9. Huang, F.; Batchelor, J.C.; Sanz-Izquierdo, B. Small size wearable button antenna. In Proceedings of the European Conference on Antennas and Propagation (EuCAP), Nice, France, 6–10 November 2006; pp. 1–4.

10. Zhu, S.; Langley, R. Dual-Band Wearable Textile Antenna on an EBG Substrate. IEEE Trans. Antennas Propag. **2009**, 57, 926–935.

11. Bashir, S. Design and Synthesis of Non Uniform High Impedance Surface based Wearable Antennas. PhD Thesis, School of Electronic, Electrical and Systems Engineering, Loughborough University, Loughborough, UK, 2009.

12. Salonen, P.; Rahmat-Samii, Y.; Schaffrath, M.; Kivikoski, M. Effect of Textile Materials on Wearable Antenna Performance: A Case Study of GPS Antenna. Int. Symp. Antennas Propag. **2004**, 1, 459–462.

13. Gupta, S.; Sankaralingam, B. Development of Textile Antennas for Body Wearable Applications and Investigations on their Performance under Bent Conditions. Progress Electromagn. Res. B **2010**, 22, 53–71.

14. Tronquo, A.; Rogier, H.; Hertleer, C.; Van Langenhove, L. Robust planar textile antenna for wireless body LANs operating in 2.45 GHz ISM band. Electron. Lett. **2006**, 42, 5–6.

15. Tronquo, A.; Rogier, H.; Hertleer, C.; Van Langenhove, L. Applying Textile Materials for the Design of Antennas for Wireless Body Area Networks. In Proceedings of the European Conference on Antennas and Propagation (EuCAP), Nice, France, 6–10 November 2006; pp. 1–5.
16. Ahmad, S.; Saidin, N.S.; Che Isa, C.M. Development of Embroidered Sierpinski Carpet Antenna. In Proceedings of the IEEE Asia-Pacific Conference on Applied Electromagnetics (APACE 2012), Melaka, Malaysia, 11–13 December 2012; pp. 123–127.
17. Sankaralingam, S.; Gupta, B. Determination of Dielectric Constant of Fabric Materials and Their Use as Substrates for Design and Development of Antennas for Wearable Applications. IEEE Trans. Instrum. Meas. **2010**, 59, 3122–3130.
18. Krupka, J.; Geyer, R.G.; Baker-Jarvis, J.; Ceremuga, J. Measurements of the Complex Permittivity of Microwave Circuit Board Substrates Using Split Dielectric Resonator and Reentrant Cavity Techniques. In Proceedings of the Seventh International Conference on Dielectric Materials Measurements & Applications, Bath, UK, 23–26 September 1996; pp. 21–24.
19. Bayram, Y.; Volakis, J.L.; Wang, Z.; Zhang, L. Multilayer printing of embroidered RF circuits on polymer composites. In Proceedings of the IEEE International Symposium on Antennas and Propagation (APSURSI 2011), Spokane, WA, USA, 3–8 July 2011; pp. 278–281.
20. Pettitt, G.; Matthews, J.C.G.; Tyler, A.J.; Pirollo, B.P. Wide-Band Body Wearable Antennas. BAE Syst. Def. Sci. Technol. Lab., IET **2008**, 111–127.
21. Waterhouse, R.; Novak, D.; Burnie, G. Small uni-planar antenna suitable for body wearable applications. In Proceedings of the Military Communications Conference, Washington, DC, USA, 23–25 October 2006; pp. 1–6.

22. Maleszka, T.; Kabacik, P. Bandwidth properties of embroidered loop antenna for wearable applications. In Proceedings of the 3rd European Wireless Technology Conference, Paris, France, 27–28 September 2010; pp. 89–92.
23. Salonen, P.; Rahmat-Samii, Y.; Hurme, H.; Kivikoski, M. Effect of conductive material on wearable antenna performance: A case study of WLAN antennas. In Proceedings of the IEEE Antennas and Propagation Society Symposium, Monterey, CA, USA, 20–25 June 2004; Volume 1, pp. 89–92.
24. Matthews, J.C.G.; Pettitt, G. Development of Flexible, Wearable Antennas. In Proceedings of the 3rd European Conference on Antennas and Propagation (EuCAP), Berlin, Germany, 23–27 March 2009; pp. 273–277.
25. Kuramoto, A. Wideband Wearable Antenna. In Proceedings of the ISAP, Niigata, Japan, 20–24 August 2007; pp. 1086–1089.
26. Salonen, P.; Hurme, H. A Novel Fabric WLAN Antenna for Wearable Applications. In Proceedings of the IEEE Antennas and Propagation Society International Symposium, Columbus, OH, USA, 22–27 June 2003; Volume 2, pp. 700–703.
27. Klemm, M.; Locher, I.; Tröster, G. A Novel Circularly Polarized Textile Antenna for Wearable Applications. In Proceedings of the 34th European Microwave Conference, Amsterdam, The Netherlands, 12–14 October 2004; Volume 1, pp. 137–140.
28. Locher, I.; Klemm, M.; Kirstein, T.; Tröster, G. Design and Characterization of Purely Textile Patch Antennas. IEEE Trans. Adv. Packag. **2006**, 29, 777–788.
29. Ouyang, Y.; Chappell, W.J. High Frequency Properties of Electro-Textiles for Wearable Antenna Applications. IEEE Trans. Antennas Propag. **2008**, 56, 381–389.
30. Ouyang, Y.; Karayianni, E.; Chappell, W.J. Effect of Fabric Patterns on Electrotextile Patch Antennas. In Proceedings of the International Symposium on Antennas and Propagation, Washington, DC, USA, 3–8 July 2005; pp. 246–249.

31. Hertleer, C.; Tronquo, A.; Rogier, H.; Vallozzi, L. Aperture-Coupled Patch Antenna for Integration into Wearable Textile Systems. Antennas Wirel. Propag. Lett. **2007**, 6, 392–395.

32. Hertleer, C.; Rogier, H.; Vallozzi, L.; Van Langenhove, L. A Textile Antenna for Off-Body Communication Integrated Into Protective Clothing for Firefighters. IEEE Trans. Antennas Propag. **2009**, 57, 919–925.

33. Wang, X.; Yao, L.; Xu, F.; Zhou, D.; Qiu, Y. Design and Characterization of Conformal Microstrip Antennas Integrated into 3D Orthogonal Woven Fabrics. J. Eng. Fibers Fabr. **2012**, 7, 76–80.

34. Acti, T.; Zhang, S.; Chauraya, A.; Whittow, W.; Seager, R.; Dias, T.; Vardaxoglou, Y. High performance Flexible Fabric Electronics for Megahertz Frequency Communications. In Proceedings of the Loughborough Antennas & Propagation Conference (LAPC), Loughborough, UK, 14–15 November 2011; pp. 1–4.

35. Zhang, S.; Chauraya, A.; Whittow, W.; Seager, R.; Acti, T.; Dias, T.; Vardaxoglou, Y. Simulation Methodology to Model the Behavior of Wearable Antennas Composed of Embroidered Conductive Threads. In Proceedings of the IEEE International Symposium on Antennas and Propagation and USNC-URSI National Radio Science Meeting, Chicago, IL, USA, 8–14 July 2012.

36. Chauraya, A.; Zhang, S.; Whittow, W.; Acti, T.; Seager, R.; Dias, T.; Vardaxoglou, Y.C. Addressing the challenges of fabricating microwave antennas using conductive threads. In Proceedings of the European Conference on Antennas and Propagation (EuCAP), Prague, Czech, 26–30 March 2012; pp. 1365–1367.

37. Seager, R.; Zhang, S.; Chauraya, A.; Whittow, W.; Vardaxoglou, Y.; Acti, T.; Dias, T. Effect of the Fabrication Parameters on the

Performance of Embroidered Antennas. IET Microw. Antennas Propag. **2013**, 7, 1174–1181.

38. Zhang, S.; Chauraya, A.; Whittow, W.; Seager, R.; Acti, T.; Dias, T.; Vardaxoglou, Y. Embroidered Wearable Antennas Using Conductive Threads with Different Stitch Spacings. In Proceedings of the Loughborough Antennas & Propagation Conference (LAPC), Loughborough, UK, 12–13 November 2012.
39. Moradi, E.; Björninen, T.; Ukkonen, L.; Rahmat-Samii, Y. Effects of Sewing Pattern on the Performance of Embroidered Dipole-Type RFID Tag Antennas. IEEE Antennas Wirel. Propag. Lett. **2012**, 11, 1482–1485. [Google Scholar] [CrossRef]
40. Koski, E.; Koski, K.; Björninen, T.; Babar, A.A.; Sydänheimo, L. Fabrication of Embroidered UHF RFID Tags. In Proceedings of the Antennas and Propagation Society International Symposium (APSURSI), Chicago, IL, USA, 8–14 July 2012; pp. 1–2.
41. Zheyu, W.; Zhang, L.; Bayram, Y.; Volakis, J.L. Embroidered Conductive Fibers on Polymer Composite for Conformal Antennas. IEEE Trans. Antennas Propag. **2012**, 60, 4141–4147.
42. Volakis, J.L.; Zhang, L.; Wang, Z.; Bayram, Y. Embroidered flexible RF electronics. In Proceedings of the IEEE International Workshop on Antenna Technology (iWAT), Tucson, AZ, USA, 5–7 March 2012; pp. 8–11.
43. Wang, Z.; Zhang, L.; Psychoudakis, D.; Volakis, J.L. Flexible textile antennas for body-worn communication. In Proceedings of the IEEE International Workshop on Antenna Technology (iWAT), Tucson, AZ, USA, 5–7 March 2012; pp. 205–208.
44. Wang, Z.; Zhang, L.; Bayram, Y.; Volakis, J.L. Embroidered E-Fiber-Polymer Composites for Conformal Load Bearing Antennas. In Proceedings of the IEEE Antennas and Propagation Society International Symposium (APSURSI), Toronto, ON, USA, 11–17 July 2010; pp. 1–4.
45. Zhang, L.; Wang, Z.; Volakis, J.L. Embroidered textile circuits for microwave devices. In Proceedings of the IEEE International

Symposium on Antennas and Propagation, Chicago, IL, USA, 8–14 July 2012; pp. 1–2.

46. Zhang, L.; Wang, Z.; Salman, S.; Volakis, J.L. Embroidered Textiles for RF Electronics and Medical Sensors. In Proceedings of the 2012 IEEE International Conference on Wireless Information Technology and Systems (ICWITS), Maui, HI, USA, 11–16 November 2012; pp. 1–4.

47. Bayram, Y.; Zhou, Y.; Shim, B.S.; Xu, S.; Zhu, J.; Kotov, N.A.; Volakis, J.L. E-Textile Conductors and Polymer Composites for Conformal Lightweight Antennas. IEEE Trans. Antennas Propag. **2010**, 58, 2732–2736.

48. Zhang, L.; Wang, Z.; Volakis, J.L. Textile Antennas and Sensors for Body-Worn Applications. IEEE Antennas Wirel. Propag. Lett. **2012**, 11, 1690–1693.

49. Osman, M.A.R.; Rahim, M.K.A.; Samsuri, N.A.; Ali, M.E. Compact and Embroidered Textile Wearable Antenna. In Proceedings of the IEEE International RF and Microwave Conference (RFM 2011), Seremban, Malaysia, 12–14 December 2011; pp. 311–314.

50. Patra, P.K.; Calvert, P.D. Textile Based Carbon Nanostructured Flexible Antenna. Natl. Text. Center Res. Briefs **2007**. NTC Project: M06-MD01.

51. Kazani, I.; Scarpello, M.L.; Hertleer, C.; Rogier, H.; de Mey, G.; Guxho, G.; Van Langenhove, L. Washable Screen Printed Textile Antennas. Adv. Sci. Technol. **2012**, 80, 118–122.

52. Kellomaki, T.; Virkki, J.; Merilampi, S.; Ukkonen, L. Towards Washable Wearable Antennas: A Comparison of Coating Materials for Screen-Printed Textile-Based UHF RFID Tags. Int. J. Antennas Propag. **2012**, 2012, 1–11. [Google Scholar]

53. Locher, I.; Troster, G. Screen-printed Textile Transmission Lines. Text. Res. J. **2007**, 77, 837–842.

54. Merilampi, S.; Björninen, T.; Haukka, V.; Ruuskanen, P.; Ukkonen, L.; Sydänheimo, L. Analysis of electrically conductive silver ink on stretchable substrates under tensile load. Microelectron. Reliab. **2010**, 50, 2001–2011.
55. Chauraya, A.; Whittow, W.G.; Vardaxoglou, J.C.; Li, Y.; Torah, R.; Yang, K.; Beeby, S.; Tudor, J. Inkjet printed dipole antennas on textiles for wearable communications. IET Microw. Antennas Propag. **2013**, 7, 760–767.
56. Whittow, W.G.; Chauraya, A.; Vardaxoglou, J.C.; Li, Y.; Torah, R.; Yang, K.; Beeby, S.; Tudor, J. Inkjet-Printed Microstrip Patch Antennas Realized on Textile for Wearable Applications. IEEE Antennas Propag. Lett. **2014**, 13, 71–74.
57. Chauraya, A.; Seager, R.; Whittow, W.; Zhang, S.; Vardaxoglou, Y. Embroidered Frequency Selective Surfaces on Textiles for Wearable Applications. In Proceedings of the Loughborough Antennas & Propagation Conference (LAPC), Loughborough, UK, 11–12 November 2013; pp. 388–391.
58. Kaufmann, T.; Fumeaux, I.; Fumeaux, C. Comparison of Fabric and Embroidered Dipole Antennas. In Proceedings of the 7th European Conference on Antennas and Propagation (EuCAP), Gothenburg, Sweden, 8–12 April 2013; pp. 3252–3255.
59. Noury, N.; Barralon, P.; Flammarion, D. Preliminary Results on the Study of Smart Wearable Antennas. In Proceedings of the 27th Annual International Conference of the IEEE Engineering in Medicine and Biology Society, Shanghai, China, 17–18 January 2005; pp. 3814–3817.
60. Salonen, P.; Sydänheimo, L.; Keskilammi, M.; Kivikoski, M. A Small Planar Inverted-F Antenna for Wearable Applications. In Proceedings of the 3rd International Symposium on Wearable Computers, San Francisco, CA, USA, 18–19 October 1999; pp. 95–100.

61. Lin, C.-H.; Li, Z.; Ito, K.; Takahashi, M.; Saito, K. Dual-mode antenna for on-/off-body communications (10 MHz/2.45 GHz). Electron. Lett. **2012**, 48, 1383–1384.
62. Huang, B.; Yan, G.; Li, Q. A four spiral slots microstrip patch antenna for radiotelemetry capsules based on FDTD. J. Zhejiang Univ. Sci. A **2007**, 8, 1560–1567.
63. Chandran, A.R.; Scanlon, W.G. Reduced Groundplane Shorted Patch Antenna for On-Body Communications. In Proceedings of the Loughborough Antennas & Propagation Conference, Loughborough, UK, 16–17 November 2009; pp. 409–412.
64. Alves, T.; Poussot, B.; Laheurte, J. PIFA—Top-Loaded-Monopole Antenna With Diversity Features for WBAN Applications. IEEE Antennas Propag. Lett. **2011**, 10, 693–696.
65. Kim, S.; Lee, S.; Kwon, K.; Park, H.; Choi, J. Miniaturized and High-Isolation Diversity Antenna for WBAN Applications. In Proceedings of the ISAP2012, Nagoya, Japan, 29 October–2 November 2012; pp. 979–982.
66. Manzari, S.; Pettinari, S.; Marrocco, G. Miniaturised wearable UHF-RFID tag with tuning capability. Electron. Lett. **2012**, 48, 1325–1326.
67. Manzari, S.; Pettinari, S.; Marrocco, G. Miniaturized and Tunable Wearable RFID Tag for Body-Centric Applications. IEEE Int. Conf. RFIDTechnol. Appl. (RFID TA) **2012**, 2, 239–243.
68. Hertleer, C.; Rogier, H.; Van Langenhove, L. A Textile Antenna for Protective Clothing. In Proceedings of the 2007 IET Seminar on Antennas and Propagation for Body-Centric Wireless Communications, London, UK, 24–24 April 2007; pp. 44–46.
69. Salonen, P.; Keskilammi, M.; Rantanen, J.; Sydanheimo, L. A novel Bluetooth antenna on flexible substrate for smart clothing. In Proceedings of the IEEE International Conference on SystemsMan and Cybernetics. e-Systems and e-Man for

Cybernetics in Cyberspace, Tucson, AZ, USA, 7–10 October 2001; pp. 789–794.

70. Salonen, P.; Sydanheimo, L. Development of an S-Band Antenna for Smart Clothing. In Proceedings of the International Symposium on Antennas and Propagation, San Antonio, TX, USA, 16–21 June 2002; pp. 6–9.

71. Carter, J.; Saberin, J.; Shah, T.; Ananthanarayanan, S.P.R. Inexpensive Fabric Antenna for Off-Body Wireless Sensor Communication. In Proceedings of the Antennas and Propagation Society International Symposium (APSURSI), Toronto, ON, USA, 11–17 July 2010; pp. 1–4.

72. Salonen, P.; Kim, J.; Rahmat-Samii, Y. Dual-Band E-Shaped Patch Wearable Textile Antenna. Antennas Propag. Soc. Int. Symp. IEEE **2005**, 1A, 466–469.

73. Salonen, P.; Rahmat-Samii, Y.; Hurme, H.; Kivikoski, M. Dual-Band Wearable Textile Antenna. IEEE Antennas Propag. Soc. Int. Symp. **2004**, 1, 463–466.

74. Kaivanto, E.K.; Berg, M.; Salonen, E.; de Maagt, P. Wearable Circularly Polarized Antenna for Personal Satellite Communication and Navigation. IEEE Trans. Antennas Propag. **2011**, 59, 4490–4496.

75. Kennedy, T.F.; Fink, P.W.; Chu, A.W.; Champagne, N.J.; Lin, G.Y.; Khayat, M.A. Body-Worn E-Textile Antennas: The Good, the Low-Mass, and the Conformal. IEEE Trans. Antennas Propag. **2009**, 57, 910–918.

76. Dobbins, J.A.; Chu, A.W.; Fink, P.W.; Kennedy, T.F.; Lin, G.Y.; Khayat, M.A.; Scully, R.C. Fabric Equiangular Spiral Antenna. In Proceedings of the IEEE Antennas and Propagation Society International Symposium, Albuquerque, NM, USA, 9–14 July 2006; pp. 2113–2116.

77. Declercq, F.; Rogier, H. Active Integrated Wearable Textile Antenna with Optimized Noise Characteristics. IEEE Trans. Antennas Propag. **2010**, 58, 3050–3054.

78. Dierck, A.; Declercq, F.; Rogier, H. Review of active textile antenna co-design and optimization strategies. In Proceedings of the IEEE International Conference on RFID-Technologies and Applications, Sitges, Barcelona, 15–16 September 2011; pp. 194–201.

79. Zhu, S.; Langley, R. Dual-band wearable antennas over EBG substrate. Electron. Lett. **2007**, 43, 141–142.

80. Salonen, P. Softwear Antenna. In Proceedings of the Military Communications Conference, San Diego, CA, USA, 16–19 November 2008; pp. 1–6.

81. Langley, R.; Zhu, S.; Liu, L. Dual Band Body Worn Antenna. In Proceedings of the European Conference on Antennas and Propagation (EuCAP), Loughborough, UK, 2–3 April 2007; pp. 137–140.

82. Chahat, N.; Zhadobov, M.; Sauleau, R.; Mahdjoubi, K. Improvement of the on-body performance of a dual-band textile antenna using an EBG structure. In Proceedings of theLoughborough Antennas & Propagation Conference (LAPC), Loughborough, UK, 8–9 November 2010; pp. 465–468.

83. Soh, P.J.; Vandenbosch, G.A.E.; Higuera-Oro, J. Design and Evaluation of Flexible CPW-fed UltraWide Band (UWB) Textile Antennas. In Proceedings of the IEEE International RF Microwave Conference (RFM 2011), Seremban, Malaysia, 12–14 December 2011; pp. 133–136.

84. Rahim, H.A.; Malek, F.; Adam, I.; Ahmad, S.; Hashim, N.B.; Hall, P.S. Design and Simulation of a Wearable Textile Monopole Antenna for Body Centric Wireless Communications. In Proceedings of the Progress in Electromagnetics Research Symposium, Moscow, Russia, 19–23 August 2012; pp. 1381–1384.

85. Moro, R.; Agneessens, S.; Rogier, H.; Bozzi, M. Wearable textile antenna in substrate integrated waveguide technology. Electron. Lett. **2012**, 48, 985–987.

86. Kaufmann, T.; Fumeaux, C. Wearable Textile Half-Mode Substrate-Integrated Cavity Antenna Using Embroidered Vias. IEEE Antennas Wirel. Propag. Lett. **2013**, 12, 805–808.

87. Seager, R.D.; Chauraya, A.; Vardaxoglou, J.C.; de Maagt, P. Towards a compact low frequency woven antenna. In Proceedings of the Antennas and Propagation Society International Symposium (APSURSI), Charleston, SC, USA, 1–5 June 2009; pp. 1–4.

88. Roh, J.-S.; Chi, Y.-S.; Lee, J.-H.; Tak, Y.; Nam, S.; Kang, T.J. Embroidered Wearable Multiresonant Folded Dipole Antenna for FM Reception. IEEE Antennas Wirel. Propag. Lett. **2010**, 9, 803–806.

89. Shobana, R.; Radha, S. Ultra Wideband Wearable Textile Antenna. In Proceedings of the International Conference on Computing and Control Engineering (ICCCR 2012), Chennai, TamilNadu, India, 12–13 April 2012; pp. 7–10.

90. Wang, Z.; Zhang, L.; Psychoudakis, D.; Volakis, J.L. GSM and Wi-Fi textile antenna for high data rate communications. In Proceedings of the 2012 IEEE International Symposium on Antennas and Propagation, Chicago, IL, USA, 8–14 July 2012; pp. 1–2.

91. Ouyang, Y.; Chappell, W.; Lafayette, W. Distributed Body-Worn Transceiver System with the use of Electrotextile Antennas. In Proceedings of the Microwave Symposium IEEE/MTT-S International, Honolulu, HI, USA, 3–8 June 2007; pp. 1229–1232.

92. Balanis, C.A. Antenna Theory Analysis and Design, 3rd ed.; John Wiley & Sons: New Jersey, NY, USA, 2005.

93. Moradi, E.; Björninen, T.; Ukkonen, L.; Rahmat-Samii, Y. Characterization of Embroidered Dipole-type RFID Tag

Antennas. In Proceedings of the IEEE International Conference on RFID-Technologies and Applications (RFID-TA), Nice, France, 5–7 November 2012; pp. 248–253.

94. Osman, M.A.R.; Rahim, M.K.A.; Azfar, M.; Samsuri, N.A.; Zubir, F.; Kamardin, K. Design, Implementation and Performance of Ultra-Wideband Textile Antenna. Progress Electromagn. Res. B **2011**, 27, 307–325.

95. Salonen, P.; Rahmat-Samii, Y. Textile Antennas: Effects of Antenna Bending on Input Matching and Impedance Bandwidth. IEEE Aerosp. Electron. Syst. Mag. **2007**, 22, 10–14.

96. Boeykens, F.; Vallozzi, L.; Rogier, H. Cylindrical Bending of Deformable Textile Rectangular Patch Antennas. Int. J. Antennas Propag. **2012**, 2012, 1–11.

97. Shimasaki, H.; Tanaka, M. Measurement of the Roundly Bending Characteristics of a Cavity Slot Antenna Made of Conductive Textile. In Proceedings of the Asia-Pacific Conference, Melbourne, Australia, 5–8 December 2011; pp. 1598–1601.

98. Tanaka, M.; Jang, J.H. Wearable Microstrip Antenna. IEEE Antennas Propag. Soc. Int. Symp. **2003**, 2, 704–707.

99. Bai, Q.; Langley, R. Crumpling of PIFA Textile Antenna. IEEE Trans. Antennas Propag. **2012**, 60, 63–70.

100. Bai, Q.; Langley, R. Crumpled textile antennas. Electron. Lett. **2009**, 45, 436–437.

101. Lilja, J.; Salonen, P.; Kaija, T.; de Maagt, P. Design and Manufacturing of Robust Textile Antennas for Harsh Environments. IEEE Trans. Antennas Propag. **2012**, 60, 4130–4140.

102. Hertleer, C.; Van Laere, A.; Rogier, H.; Van Langenhove, L. Influence of Relative Humidity on Textile Antenna Performance. Text. Res. J. **2009**, 80, 177–183.

103. Koski, K.; Moradi, E.; Babar, A.A.; Björninen, T.; Sydänheimo, L.; Ukkonen, L. Durability of Embroidered Antennas in Wireless

Body-Centric Healthcare Applications. In Proceedings of the 7th European Conference on Antennas and Propagation (EuCAP), Gothenburg, Sweden, 8–12 April 2013; pp. 565–569.

104. Zhang, S.; Chauraya, A.; Whittow, W.; Seager, R.; Acti, T.; Dias, T.; Vardaxoglou, Y. Repeatability of Embroidered Patch Antennas. In Proceedings of the Loughborough Antennas & Propagation Conference (LAPC), Loughborough, UK, 11–12 November 2013; pp. 140–144.

105. Koski, K.; Koski, E.; Björninen, T.; Babar, A.; Ukkonen, L.; Sydänheimo, L. Practical Read Range Evaluation of Wearable Embroidered UHF RFID Tag. In Proceedings of the IEEE Antennas and Propagation Society International Symposium (APS-URSI), Chicago, IL, USA, 8–14 July 2012; pp. 1–2.

106. Boyes, S.J.; Soh, P.J.; Huang, Y.; Vandenbosch, G.A.E.; Khiabani, N. Measurement and Performance of Textile Antenna Efficiency on a Human Body in a Reverberation Chamber. IEEE Trans. Antennas Propag. **2013**, 61, 871–881.

107. Psychoudakis, D.; Lee, G.; Chen, C.; Volakis, J.L. Military UHF Body-Worn Antennas for Armored Vests. In Proceedings of the European Conference on Antennas and Propagation (EuCAP), Barcelona, Spain, 12–16 April 2010; pp. 1–4.

108. Zhang, L.; Wang, Z.; Psychoudakis, D.; Volakis, J.L. E-fiber electronics for body-worn devices. In Proceedings of the 6th European Conference on Antennas and Propagation (EUCAP), Prague, Czech, 26–30 March 2012; pp. 760–761.

109. Tsolis, A.; Whittow, W.G.; Alexandridis, A.A.; Vardaxoglou, J.C. Evaluation of a Human Body Phantom for Wearable Antenna Measurements at the 5.8 GHz Band. In Proceedings of the Loughborough Antennas & Propagation Conference (LAPC), Loughborough, UK, 11–12 November 2013; pp. 414–419.

110. Santas, J.G.; Alomainy, A. Textile antennas for on-body communications: Techniques and properties. In Proceedings of

the European Conference on Antennas and Propagation (EuCAP), Edinburgh, Scotland, 11–16 November 2007; pp. 1–4.

111. Hasani, M.; Vena, A.; Sydänheimo, L.; Ukkonen, L.; Tentzeris, M.M. Implementation of a Dual-Interrogation-Mode Embroidered RFID-Enabled Strain Sensor. IEEE Antennas Wirel. Propag. Lett. **2013**, 12, 1272–1275.

112. Raad, H.R.; Abbosh, A.I.; Al-Rizzo, H.M.; Rucker, D.G. Flexible and Compact AMC Based Antenna for Telemedicine Applications. IEEE Trans. Antennas Propag. **2013**, 61, 524–531.

113. Jais, M.I.; Jamlos, M.F.B.; Jusoh, M.; Sabapathy, T.; Kamarudin, M.R.; Ahmad, R.B.; Azremi, A.A.A.-H.; Bin Azmi, E.I.; Soh, P.J.; Vandenbosch, G.A.E.; et al. A Novel 2.45GHz Switchable Beam Textile Antenna (SBTA) for Outdoor Wireless Body Area Network (WBAN) Applications. Progress Electromagn. Res. **2013**, 138, 613–627.

114. Sanchez-Hernandez, D.A. High Frequency Electromagnetic Dosimetry; Artech House: London, UK; Boston, MA, USA, 2009.

115. Seager, R.D.; Chauraya, A.; Zhang, S; Whittow, W.; Vardaxoglou, Y. Flexible radio frequency connectors for textile electronics. Electron. Lett. **2013**, 49, 1371–1373.

116. Global smartphone shipments top one billion 'milestone'. Available online: http://www.bbc.co.uk/news/business-25923404 (accessed on 27 May 2014).

117. The Internet of Things. Available online: http://share.cisco.com/internet-of-things.html (accessed on 27 May 2014).

118. CES 2013: Ready for the wearable tech revolution. Available online: http://www.bbc.co.uk/news/technology-20956970 (accessed on 27 May 2014).

119. Wearable technology market to exceed $6B by 2016. Available online: http://www.computerworld.com/s/article/9230095/Wearable_tec

hnology_market_to_exceed_6B_by_2016?goback=%2Egde_73268_member_146470831 (accessed on 27 May 2014).

120. Wearable tech revenue to hit $19B by 2018. Available online: http://www.cnet.com/uk/news/wearable-tech-revenue-to-hit-19b-by-2018/ (accessed on 27 May 2014).

121. Worldwide Annual Shipments of Smartwatches 2012-2020. Available online: http://boygeniusreport.files.wordpress.com/2013/09/smartwatch-shipments.png (accessed on 27 May 2014).

122. Cihangir, A.; Whittow, W.G.; Panagamuwa, C.J.; Ferrero, F.; Jacquemod, G.; Gianesello, F.; Luxey, C. Feasibility Study of 4G Cellular Antennas for Eyewear Communicating Devices. Antennas Wirel. Propag. Lett. IEEE **2013**, 12, 1704–1707.

123. Microsoft bought wearable tech assets for $150m. Available online: http://www.develop-online.net/news/microsoft-bought-wearable-tech-assets-for-150m/0191147 (accessed on 27 May 2014).

124. 10m alive in Britain today will live to be more than 100 years old. Available online: http://www.telegraph.co.uk/health/8230749/10m-alive-in-Britain-today-will-live-to-be-more-than-100-years-old.html (accessed on 27 May 2014).

125. Kellomaki, T.; Whittow, W.G.; Heikkinen, J.; Kettunen, L. 2.45 GHz Plaster Antennas for Health Monitoring. In Proceedings of the 3rd European Conference on Antennas and Propagation (EuCAP), Berlin, Germany, 23–27 March 2009; pp. 211–215.

126. Powerweave: Development of textiles for electrical energy generation and storage. Available online: http://www.powerweave.eu (accessed on 27 May 2014).

CHAPTER 6

THE CHALLENGE OF PRODUCING FIBER-BASED ORGANIC ELECTRONIC DEVICES

Tobias Könyves-Toth, Andrea Gassmann and Heinz von Seggern *

Electronic Materials Division, Institute of Materials Science, Technische Universität Darmstadt, Alarich-Weiss-Str. 2, 64287 Darmstadt, Germany

ABSTRACT

The implementation of organic electronic devices on fibers is a challenging task, not yet investigated in detail. As was shown earlier, a direct transition from a flat device structure to a fiber substrate is in principle possible. However, a more detailed investigation of the process reveals additional complexities than just the transition in geometry. It will be shown, that the layer formation of evaporated materials behaves differently due to the multi-angled incidence on the fibers surface. In order to achieve homogenous layers the evaporation process has to be adapted. Additionally, the fiber geometry itself facilitates damaging of its surface due to mechanical impact and leads to a high surface roughness, thereby often hindering commercial fibers to be used as substrates. In this article, a treatment of commercial polymer-coated glass fibers will be demonstrated that allows for the fabrication of rather flexible organic light-emitting diodes (OLEDs) with cylindrical emission characteristics.

Since OLEDs rely the most on a smooth substrate, fibers undergoing the proposed treatment are applicable for other organic electronic devices such as transistors and solar cells. Finally, the technique also supports the future fabrication of organic electronics not only in smart textiles and woven electronics but also in bent surfaces, which opens a wide range of applications.

KEYWORDS

organic electronics; fibers; smart textiles; organic light emitting diodes; thin film deposition

1. INTRODUCTION

The vision of mass-produced smart textiles with electronic functionalities such as light emission or current switching for applications in textile displays, flexible lighting and wearable electronics is currently an interesting topic in textile research. In order to achieve this aim, a simple production and integration of semiconducting devices into textiles is in the focus of current research. Two different approaches are investigated presently, firstly, the integration of classical, inorganic electronic devices into fabrics [1] and, secondly, the development of new fiber-based device structures utilizing textile substrates [2]. Despite of their advantage of well-known electrical behavior and good performance, inorganic electronic devices have the disadvantage of being hard and rigid. Consequently, they oppose the main textile properties of being soft and flexible. In order to maintain these properties, the research of new OLED device structures assesses the possibility of integration without destroying the desired textile properties, however, ensuring economic production [1,2]. In this respect, organic electronics is an interesting alternative due to its thin layer structures (about 100 nm layer thickness) and their flexibility. Since most textiles consist of fabric woven or knitted from yarn, the possibilities of applying organic semiconductor devices directly onto fibers have been investigated. Reports from literature range from organic light-emitting diodes (OLEDs) [2,3] over organic field-effect transistors (OFETs) [4,5,6,7,8,9] to organic photovoltaic cells (OPV) [10,11,12,13,14]. However, only a few of the published devices cover the complete circumference of 360° of the fiber [2,3,10,11,12]. This becomes important if the size of the active area is of concern like in OLEDs or sensing applications. In principle, such fiber-based devices could be directly woven into the fabric and contacted in the same production step by conductive yarns [15,16], thereby opening the possibility of a potentially low cost production of textile-integrated electronics.

In their paper from 2007, O'Connor and coworkers [3] showed the working principle of a fiber-based small molecule OLED and compared it to a device in flat geometry. However, since that time nothing alike or new was reported for small molecule-based OLEDs on fibers. Consequently, the question arises what actually hampers further progress in fiber shaped small molecule-based organic devices. To answer this question a more detailed look on the processing of devices on fiber substrates is taken and will detail the processing-related challenges in this field. For these investigations it is mandatory to be able to reproducibly prepare rotationally symmetric devices. Thus, the first step was to develop a vacuum deposition system capable of meeting the necessary requirements for fiber substrates. In the custom-made system all process steps up to the complete device are executed automatically, starting from cleaned substrates. In the second step, the handling of the substrates during device preparation had to be elaborated and the layer formation was investigated separately. Finally, top emitting OLEDs were produced using a prototypical material stack [17] with process parameters adapted from flat geometry devices. In the course of these experiments critical parameters and material-related issues could be identified which will be discussed in detail. Especially the importance of the fiber substrate quality will be addressed and the question how to obtain smooth surfaces on commercially available fibers by a thermal treatment will be investigated.

2. RESULTS AND DISCUSSION

2.1. Effect of the Transition from Flat to Cylindrical Substrate Geometry on Layer

A reliable automated vacuum deposition system is essential for the investigation of a production process and its variation. In order to be able to fulfill all necessary requirements for fibers as substrates, a system was built from scratch. It was designed to fit into a vacuum deposition chamber (Auto 306 system from HHV Ltd., Crawley, UK) offering complete automation of device structuring by shadow masks and rotation of the fibers (see experimental section for a detailed description). Figure 1 shows a photographic image of the described system before its integration into the vacuum chamber.

For the investigation of possible necessary process alterations when changing from flat device structures to fibers, OLEDs instead of other devices like OFETs, etc., were chosen as model devices. They exhibit a simple device structure and, additionally to electrical characterization, can also be evaluated optically. Furthermore, they have the highest requirements on the substrate surface quality making the results thus applicable to all other organic device structures. Due to their small layer thickness, organic semiconductor devices are in general very sensitive to the surface roughness of the substrate. To reliably fabricate functional OLEDs usually a root mean square (RMS) roughness below 1 nm is required [18]. In addition, for future textile applications mechanically flexible polymer fibers are mostly suited, however, problems in the surface quality have to be overcome. Therefore, initially rigid glass sticks with a diameter of 1 mm were used to allow for an easier substrate handling and to understand the influence of the cylindrical geometry on layer formation. Firstly, only the geometry was changed from a flat glass substrate to a cylindrical glass rod. To make sure that the glass surface of the substrates was not damaged, they were fabricated in house by glassblowers and stored without coming in contact to other surfaces.

Figure 1. Photographic image of the developed automated system integrated into an evaporation chamber for the production of fiber-based organic electronic devices.

It was found that basically all control parameters used for flat glass substrates, namely vacuum pressure, deposition rate and substrate temperature can be directly used for the deposition on rotated fibers. Only for calculation of the layer thickness, which is usually calibrated for a flat reference substrate, the ratio of circumference to diameter π has to be taken into account. In addition, it was discovered that the rotation speed of the fiber substrates is an additional control parameter, which has to be adjusted.

It is known, that the layer growth and morphology of thin films depends for oblique evaporation on the incident angle [19]. By changing the angle of incidence during layer deposition it is thus possible to change the properties of the layer. For a fiber geometry, multiple angles are present at the same time. This results in an even more complex layer growth, which was found to be directly affected by the rotation speed of the fibers. In addition experiments require a poly(3,4-ethylenedioxythiophene) poly(styrene-sulfonate) (PEDOT:PSS) coating of the substrates as is explained below. It acts as adhesive and nucleation layer for the silver layer used as underlying layer for the anode [20]. Figure 2 shows the morphology of silver layers deposited at different rotation speeds on PEDOT:PSS-coated glass sticks. The layers produced on non-rotated fibers (Figure 2a) and on at a rotation speed of 4 rpm rotated fibers (Figure 2b) show an incomplete surface coverage, leading to cracks in the on top evaporated gold layer that was used as anode in the OLEDs (Figure 2d). These cracks resulted in a fast burn-out of the respective OLEDs. However, using a fiber rotation speed of 0.5 rpm a complete coverage of the PEDOT:PSS-coated glass stick by the silver layer could be achieved (Figure 2c), leading to a considerable reduction in the number and size of cracks in the Au layer and to more stable OLED devices (see Figure 2e). O'Connor et al. [3] utilized 30 rpm on a polyimide coated fiber and thus demonstrated that faster rotation can also yield good results. Although we do not yet understand the mechanism why, in our case, a slow fiber rotation is superior to a faster one, the adhesion of the metal on the glass surface plays an important role as stated above. Additionally silver layers without an underlying PEDOT:PSS layer always showed an incomplete coverage (comparable to Figure 2b) independent of the used rotation speed. Consequently devices without PEDOT:PSS were not functional due to immediate burn-out like the devices with polymer coating and a rotation speed of 4 rpm.

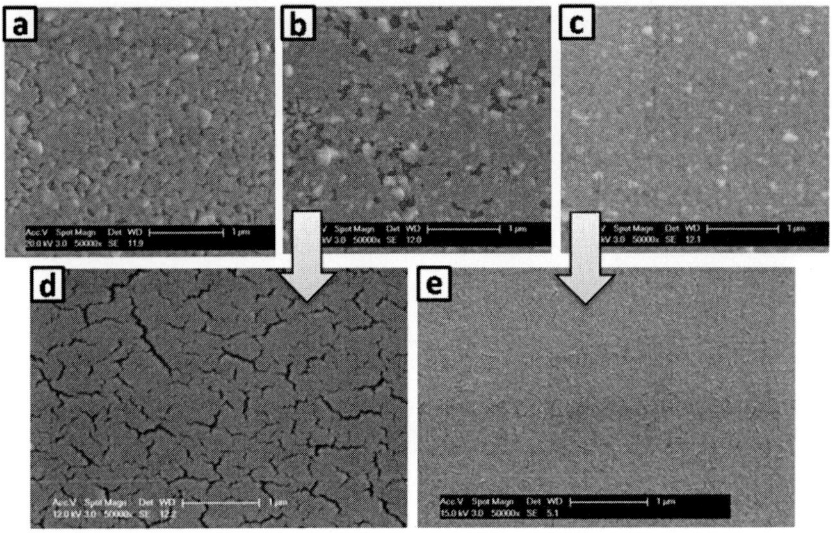

Figure 2. Scanning electron microscope images of Ag and Au surfaces. All layers were deposited with varying rotation speeds onto glass sticks (diameter 1 mm) coated with 110 nm PEDOT:PSS. (**a**) 20 nm Ag (no rotation); (**b**) 20 nm Ag (4 rpm); (**c**) 20 nm Ag (0.5 rpm); (**d**) 30 nm Au (4 rpm) on 20 nm Ag (4 rpm); (**e**) 30 nm Au (4 rpm) on 20 nm Ag (0.5 rpm).

The layer formation of the first metallic layers deposited on the substrate is crucial for the functionality of the final OLED due to the above-mentioned requirements. In contrast to the relevance of the layer quality of the polycrystalline metallic layers, the layer formation of the organic small molecules is not an issue. The prototypical organic semiconductors used are known to form amorphous layers with a smooth surface [21,22].

2.2. Electrical and Optical Device Properties

After having evaluated the effect of the substrate rotation speed on the formation of the metallic layers used as anode, small molecule OLEDs were prepared and electro-optically characterized (see Section 3 for experimental and device structure details). Figure 3 shows the current density j as a function of the electric field F of an OLED on a glass stick (1 mm diameter) with 1 mm length, which was measured in dry nitrogen atmosphere using a HP 4155A semiconductor parameter analyzer and a photodiode (BPW 34, Osram, Regensburg, Germany). It has to be noted that the electric field corresponds to voltages of up to 21 V and is thus higher than typically applied to flat OLEDs using the same organic semiconductors [17]. This mainly results from the voltage drop along the cathode which on the one hand has to be kept thin to out-couple the generated light (see Figure 4) but resulting on the other hand in a high resistivity. This influence can be reduced if the device is contacted directly in the region of the active area, however this was avoided to minimize the risk of damaging the layers by the contact needles.

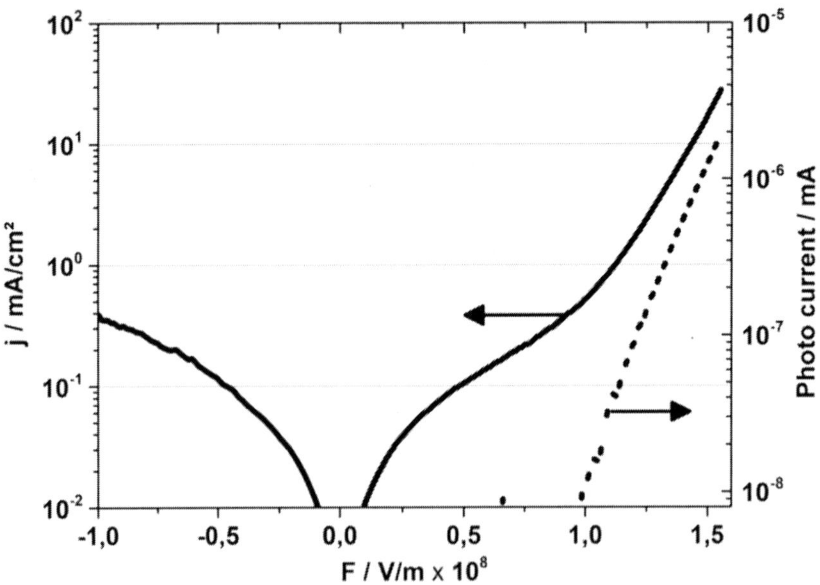

Figure 3. Electrical characteristics of a 1 mm long OLED processed on a defect-free glass stick (diameter 1 mm).

The angle dependence of the light emission was also measured using an ensemble of three photodiodes (Siemens BPW 34B, Munich, Germany). They were moved at a fixed radius and at an angle of 45° to each other around the OLED, as is depicted in Figure 4a. Consequently the diodes are covering a total angle of up to 250° around the circumference and verifying each other's results by the overlapping range. A coverage of the whole 360° circumference was not possible due to the mounting and contacting of the fiber. The result is plotted in Figure 4b and shows a homogenous emission in all directions. The measured higher emission intensities at low angles (below 90°) are related to a slightly non-centric positioning of the OLED in the setup and do not represent a real local increase in emission intensity. The sudden decrease in the signal of photodiode 2 at about 95° is attributed to a loss of contact at this angle. Since the photodiodes 1 and 3 stay at level, they verify the result. The same can be seen in the signal of photodiode 1 between 0° and 60°. In Figure 4 the photographs of the measured OLED in the on and off state are given. From the bottom picture of Figure 4c one can clearly see that the cathode stack is semitransparent. At the electroluminescence maximum (530 nm for Alq_3), the transmission loss is about 50% for the used cathode layer thickness. Even though the photograph of the switched on OLED seems to emit rather bluish, the measured electroluminescence spectrum in Figure 4d proves that the emission maximum is indeed in the green wavelength range as expected from Alq_3. The seemingly observed color change is actually an artifact of the camera, due to a non-adjusted white balance. Nevertheless, the photo nicely illustrates the homogeneous light emission of the device.

2.3. Smooth Surfaces on Commercial Fibers

Even though the well-performing OLEDs on defect-free glass sticks proved that these substrates are in general suitable for organic devices, the approach is not applicable for smart textiles. The reason is that a sufficient surface quality is in general not given for commercial fibers as will be detailed below. Thus, the obtained processing technology for cylindrical substrates has to be implemented for mechanically flexible, commercially available optical fibers. Thereby, the successful preparation of fiber-based devices shifts back to the question of the surface quality of the fiber itself and to means to smoothen it.

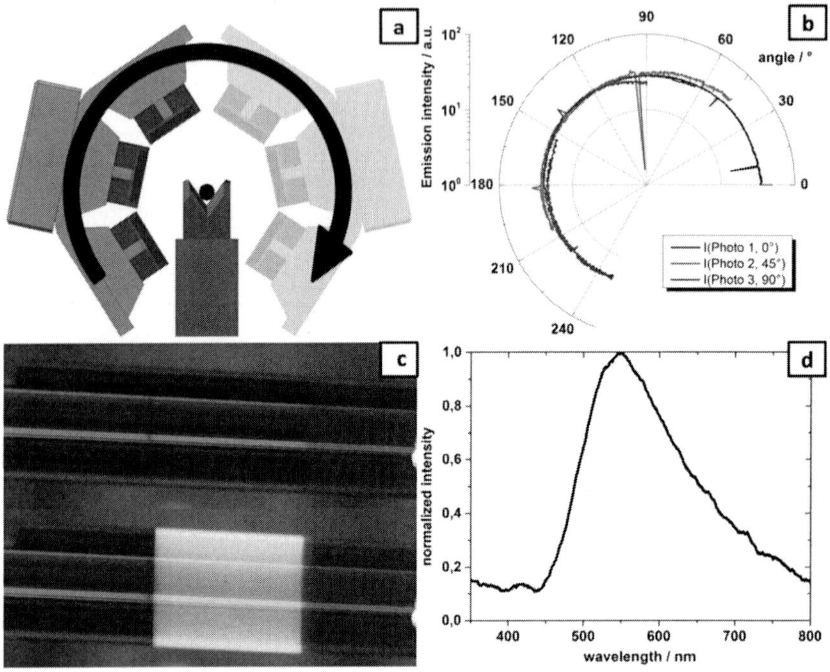

Figure 4. (**a**) Schematic of the measurements setup for angle dependant emission. The fiber (black, **center**) is contacted by laying in contact notches and the ensemble of photodiodes (blue) moved on a fixed radius around the OLED; (**b**) Angle dependence of light emission of an OLED processed on a defect-free glass stick using three photodiodes that were rotated around the device; (**c**) Photographs of the 1 mm OLED in the off (**top**) and on state (**bottom**); (**d**) Electroluminescence spectra of the diode.

In general, the common way to get smooth (flat) substrates is the fabrication of the surface from molten materials, since surface tension tries to minimize the surface energy. Like the common float glass, these substrates have a very low surface roughness. Basically, the same holds true for fibers spun from the melt. However, contrary to flat glass substrates, the smooth surface is not easily maintained in the case of fibers. The storage of fibers (and yarns) usually includes a contact of the fiber surface to other objects, leading to a damage of the surface by mechanical scratches. The obvious example is the yarn on a bobbin scratching itself while being coiled up. The same holds true for the transport of the yarn by guiding systems in the production process. To protect the optical fiber from such damages a buffer coating or cladding is usually applied. In the case of glass fibers this coating also provides the mechanical integrity of the whole fiber.

Since, in general, no fresh and, thus, undamaged optical fibers are available, a re-melting of the surface could heal failures. However, due to their low ratio of volume to surface a sole melting of the surface is difficult to achieve. Thus, unclad fibers will lose their form and/or diameter when heated above their melting temperature. In this respect, glass fibers coated with a buffer layer may have an advantage. If treated properly the surface of a polymeric buffer layer can be melted while the actual glass fiber does not lose its cross sectional shape and its diameter due to its high melting point which keeps the surface of the cladding in its form.

To prove the above-described principle of a temperature treatment hard-clad silica (HCS) optical fibers encased in a buffer layer of ethylene tetrafluoroethylene (ETFE) with two different diameters were utilized. The thicker fibers with an outer diameter of 1040 μm had a HCS core diameter of 630 μm and a buffer layer thickness of 205 μm. The thinner fibers with an outer diameter of 500 μm exhibited a HCS core of 230 μm in diameter and the ETFE buffer was 135 μm thick. For the investigation of the as received fiber surface and the ETFE buffer roughness, a 3D laser scanning microscope (Confocal Laser Scanning Microscope VK9700, KEYENCE, Neu-Isenburg, Germany) was used allowing for the recording of the radial surface roughness. The image of the surface of the 1040 μm thick, commercial HCS fiber is depicted in Figure 5a and shows various deep scratches. The peak-to-peak roughness reaches values of about 42 μm, making this fiber inapplicable as OLED substrate. The melting temperature of ETFE was measured by differential scanning calorimetry (DSC 200 F3 Maia®, Netzsch Gerätebau GmbH, Selb, Germany) to be 250 °C in agreement with literature [23]. A temperature treatment of the fibers in an oven at a temperature exceeding the melting point of ETFE results in the desired smooth surface (see Section 3 for experimental details). In Figure 5, the image of the fiber surface and its height profile after thermal treatment are depicted in comparison to the original state. It should be noted that the same part of the fiber is shown. The resulting height profile displays a roughness below the resolution limit of the laser scanning microscope. The investigation of the surface by atomic force microscopy (AFM) is included in Figure 5 and yields an RMS surface roughness of 11 nm, which is consistent with published values for ETFE [24]. Even though the roughness of the surface is still higher than recommended for OLEDs (r(RMS) < 1 nm), these substrates can already be successfully used as will be demonstrated in the following. In principle the described smoothing process should be applicable to every material combination as long as the core of the fiber has a sufficiently higher melting point than the mantle material.

2.4. OLEDs on Thermally Smoothened HCS Fibers

The j-F-characteristics of OLEDs processed on thermally smoothened ETFE-clad optical fibers with different diameters are shown in Figure 6. One OLED is 1 mm long and has been processed on a 1040 μm thick fiber, thus having an active area of 3.3 mm^2. The other OLED has been prepared on a thinner fiber of 500 μm diameter, using a larger device length of 5 mm yielding an active area of 7.9 mm^2. For measuring the light emission the used photo diodes were not sensitive enough. The photographs shown in Figure 6 showing the OLEDs in on state were taken at exposure times of up to 120 s to capture the working diode. As can be deduced from Figure 6b the fibers are already quite flexible and even for bend devices the OLEDs are still functioning. However, the j-F-characteristics of the devices processed on the thinner fibers show some artifacts in the range from 40 MV/m to 90 MV/m resembling a negative differential resistance, which could result from the high surface roughness of 11 nm of the thermally treated ETFE and therewith of the electrode interfaces [25]. Furthermore, the devices are not yet stable to withstand repeated or alternated bending, yielding currently an electrical device failure by a short circuit after 10 bending cycles to a radius of 10 mm. The reason for the failure is yet to be determined. Due to the failure by electric short circuit we assume it to be related to a crack propagation of the still existing cracks in the anode (compare Figure 2e) leading to a higher charge injection at the cracks and thus burn out of the devices.

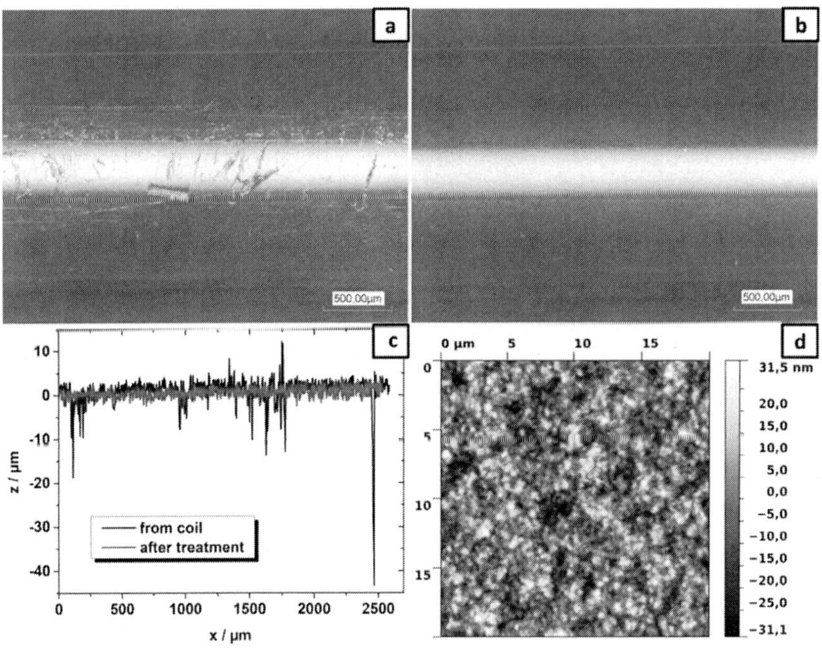

Figure 5. (**a**) 3D laser scanning microscope images of the same ETFE-coated optical fiber before (from coil) and (**b**) after the temperature treatment; (**c**) The surface profile shows that the peak-to-peak roughness is reduced from about 50 µm to a value lower than the resolution of the imaging method; (**d**) AFM image of the ETFE surface after treatment. The resulting RMS surface roughness is 11 nm.

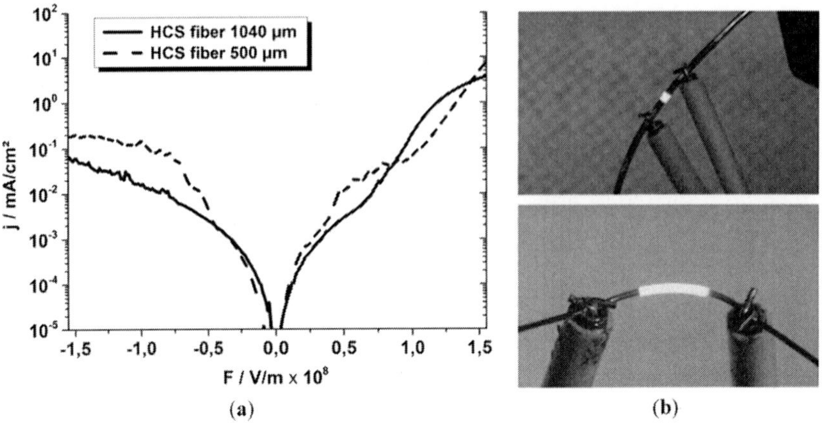

Figure 6. (**a**) j-F-characteristics and (**b**) photographs of a 1 mm OLED produced on a 1040 μm thick ETFE-coated, thermally smoothed HCS fiber (**top**) and a 5 mm OLED produced on a 500 μm thick HCS fiber treated the same way (**bottom**).

2.5. Other Routes to Improve the Surface Quality of All-Polymeric Fibers

To even further reduce the surface roughness other smoothing techniques were investigated. In the production of OLEDs a remaining surface roughness is usually treated by the application of a second layer in order to smoothen it. A prominent example for this technique is the PEDOT:PSS-coating of transparent conductive oxides like indium tin oxide (ITO) [18]. A prerequisite for this technique is that the employed solution or dispersion of the material to be coated shows a good wetting of the substrate surface. This is, unfortunately, not the case for the above discussed ETFE-clad glass fibers due to the high surface tension [24]. Consequently, all polymeric optical fibers were chosen in order to exploit other smoothing strategies, even though they were unsuited as OLED substrates. Due to the lack of the high melting core the thermal smoothing of the scratched surface was not applicable for these fibers and the surface consequently too rough to support the thin device structures.

Investigating as received, unclad polymeric fibers the areas between the deep scratches showed actually a surface roughness in the range of the heat-treated ETFE cladding on HCS fibers. In these regions it is legitimate to assume that comparable devices to those of the heat-treated glass fibers with ETFE-cladding can be obtained. In order to improve the surface roughness additional coatings can be applied as mentioned above. In the case of poly(methyl methacrylate) (PMMA) fibers it could be demonstrated that an additional coating with two different polymer layers, consisting of polyvinyl alcohol (PVA) and PMMA reduced the surface roughness below 1 nm. During the processing the use of orthogonal solvents is mandatory in order to not dissolve the fiber itself, nor the first coated polymer layer. To smoothen the PMMA fibers (diameter 500 μm) both layers were applied from solution using dip coating (see Section 3 for experimental details).

The fibers showed an initial RMS surface roughness of 5 nm as imaged by AFM (Figure 7a). Applying the PVA layer results in a reduction of the roughness to 2.4 nm, about half of the initial value (Figure 7b). After the second, subsequently applied PMMA coating the final roughness was measured to be 0.9 nm (Figure 7c) and, thus, meets the initially discussed requirements for OLED applications. These results suggest that a fiber with a high melting core encased in a mantle of PMMA should be perfectly suitable for OLEDs after the described thermal smoothing and subsequent coating with polymer solutions.

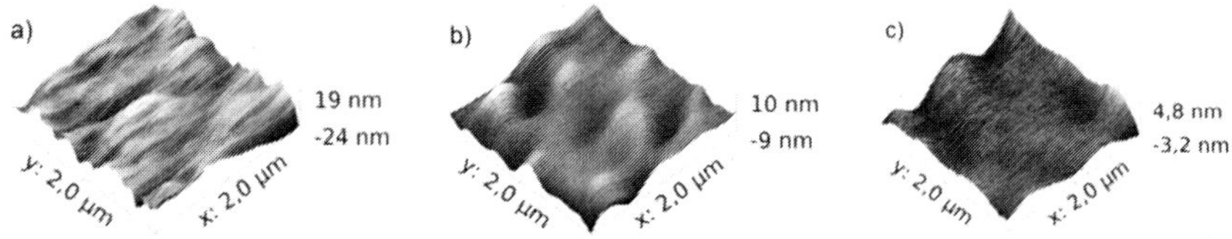

Figure 7. The surface of a PMMA fiber as imaged by AFM. The corresponding RMS roughness is given in brackets: (**a**) Untreated fiber (5 nm); (**b**) fiber after coating with PVA (2.4 nm); (**c**) fiber after subsequent coating with PVA and PMMA (0.9 nm).

These findings can be utilized to further speed-up the development and application of fiber-based devices. Since methods to control the fiber substrate surface quality and the importance of the adjustment of the rotation speed during the layer deposition have been demonstrated, the focus of future experiments can be directed to the development of concepts for a continuous manufacturing process and to exploit methods to contact functional devices on fibers by conductive yarns. There is still a long way to go to the integration of organic semiconductor-based devices into textile applications, yet first necessary steps have been demonstrated.

3. EXPERIMENTAL SECTION

Prior to the deposition of the active layers all fiber substrates were cleaned by a sequence of washing steps in an ultrasonic bath. First, the substrates were rinsed in deionized water (DI-water) and subsequently washed in a solution of the detergent Deconex (Deconex PF15, 5 wt% in DI-water, Borer Chemie AG, Zuchwil, Switzerland) at 60 °C for 15 min. In the second step, they were washed in DI-water at room temperature for 15 min to remove all excess Deconex solution left on the fibers. In the last step the fibers were washed in 2-propanol (VLSI Selectipur®, BASF, Rhine, Germany) at room temperature for 15 min. After drying the substrates in a nitrogen stream they were ready to be coated with the adhesion promoting layer of PEDOT:PSS (Clevios AI4083 aqueous dispersion, Heraeus, Hanau, Germany). This is done by dipping the fibers into the PEDOT:PSS-dispersion and withdrawing them at a speed of 80 mm/min, resulting in a layer thickness of 110 nm. Afterwards, the substrates are mounted in the sample holder to be transferred into the deposition chamber.

In the vacuum deposition system the sample holder for the fiber substrates can be decoupled from the fiber's rotation drive to allow the sample transfer via a load lock in inert nitrogen atmosphere. This is important especially after the process, since the small molecule-based organic semiconductors are in general sensitive to moisture and oxygen. The sample holder can hold up to 5 fibers with a length of 70 mm each, of which 50 mm can be structured. Mounted fibers can be rotated centric and homogeneously at speeds of up to 5 rpm. For the structuring of the OLED layers the system can in situ access 4 shadow masks. These masks are mounted in a magazine and can be positioned under the rotating fibers, if required. Monitoring quartz sensors allow the system to be controlled via a computer interface. To fully utilize the capability of the system the evaporation sources of the vacuum chamber were modified to hold up to 8 resistively heated crucibles. The active material source can be switched in situ without venting the vacuum chamber, enabling the system to hold the base pressure below 10^{-6} mbar during the whole evaporation process. The system is completed by a home-made transfer system that can be attached to the chamber thereby allowing for a sample transfer to the glove box where the electrical measurements are performed.

Since gold is easily scratched off the glass surface, a better adhesion and contacting of the Au anode was achieved by depositing an underlying layer of 20 ± 2 nm Ag before 30 ± 1 nm Au were deposited. The anode was followed by a double layer of organic semiconductors. N,N′-Di-[(1-naphthyl)-N,N′-diphenyl]-1,1′-biphenyl)-4,4′-diamine (α-NPD) served as hole transport layer and tris-(8-hydroxyquinoline) aluminum (Alq_3) was used as emitter and electron transport layer. Both layers had a thickness of 70 ± 1 nm, respectively. Finally, 0.7 ± 0.05 nm of LiF, 9 ± 1 nm of Al and 20 ± 2 nm of Ag were deposited as a semitransparent cathode stack. All materials were deposited by resistive evaporation in a vacuum chamber utilizing deposition rates of 1–2 Ås^{-1} (0.1 Ås^{-1} for LiF) and pressures of 5×10^{-7} to 3×10^{-6} mbar. The deposition was only stopped at full rotations in order to get a homogeneous layer thickness over the whole circumference. Thus variations of the deposition rate resulted in the above-mentioned variations in the layer thickness since the deposition was stopped e.g., one rotation earlier or later. With the exception of the Ag layer, all other layers were deposited at a rotation speed of 4 rpm. For the Ag layers underlying the anode and the layer covering the semitransparent cathode 0.5 rpm and 2 rpm were used, respectively. Per substrate, 3 to 5 OLED devices were structured using shadow masks. The overlapping electrodes resulted in OLEDs of 1 mm or 5 mm length. The substrates where not heated. Over the whole process their temperature never exceeded 60 °C. Figure 8 illustrates the described device structure.

For the smoothing of the ETFE-clad HCS fibers, they were cleaned properly prior to annealing in order to avoid any partial embedment of, e.g., dust particles in the molten buffer or the creation of holes in the surface during the exposure to high temperatures. The same cleaning as described above was used. Afterward the fibers were annealed at 280 °C for 15 min. and after removing from the oven cooled to room temperature. The following evaporation of the OLED was executed without additional cleaning.

For investigation of the polymer smoothing the PMMA fiber was cleaned as above and subsequently coated by dip coating. In order to not dissolve the PMMA fiber surface, an aqueous solution of polyvinyl alcohol (PVA from Sigma Aldrich, Steinheim, Germany; 3 wt% in DI-water) was dip coated first and the fiber was subsequently dried for 5 min at 105 °C in an oven. In the second step the fiber was coated using a PMMA-solution (PMMA from Sigma Aldrich, Steinheim, Germany; 6 wt% in toluene). Again the fiber was dried for 5 min at 105 °C in an oven. Both coatings where applied using a dip coating speed of 80 mm/min.

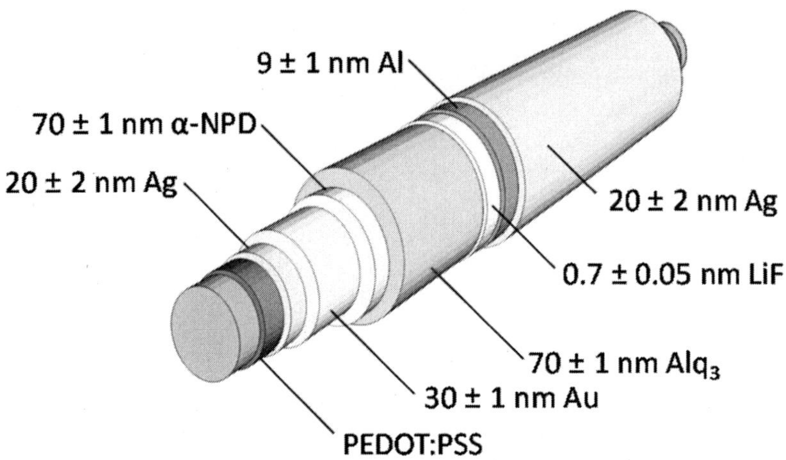

Figure 8. Illustration of the fiber OLED layer sequence. All materials were deposited by subsequent resistive evaporation in a vacuum chamber.

ACKNOWLEDGMENTS

Financial support from the Federal Ministry of Education and Research of Germany for the LUMOLED project (BMBF #16SV4040) is acknowledged. The authors want to thank R. Stark and M. Janko from the Center of Smart Interfaces at TU Darmstadt, Physics of Surfaces Department for the use of the 3D laser scanning microscope and L.-O. Heim from the Center of Smart Interfaces at TU Darmstadt, Experimental Interface Physics Department for the use of the AFM.

AUTHOR CONTRIBUTIONS

T.K.-T. has designed the setup and has performed the experimental work in the framework of a doctoral thesis. Naturally he is earning the most credit for the present publication. A.G. is the acting supervisor and discussion partner for everyday questions and problems related to experimental and processing questions. H.V.S. is the responsible project leader who has determined the overall direction of the work. He has given valuable input to the success of the work.

REFERENCES

1. Cherenack, K.; van Pieterson, L. Smart textiles: Challenges and opportunities. J. Appl. Phys. **2012**, 112, 091301:1–091301:14.

2. Janietz, S.; Gruber, B.; Schattauer, S.; Schulze, K. Integration of OLEDs in Textiles. Adv. Sci. Technol. **2013**, 80, 14–21.

3. O'Connor, B.; An, K.H.; Zhao, Y.; Pipe, K.P.; Shtein, M. Fiber shaped light emitting device. Adv. Mater. **2007**, 19, 3897–3900.

4. Lee, J.B.; Subramanian, V. Organic transistors on fiber: A first step towards electronic textiles. In Proceedings of the IEEE International Electron Devices Meeting, 2003 (IEDM'03 Technical Digest), Washington, DC, USA, 8–10 December 2003.

5. Lee, J.B.; Subramanian, V. Weave patterned organic transistors on fiber for E-textiles. IEEE Trans. Electron Devices **2005**, 52, 269–275.

6. Bonfiglio, A.; de Rossi, D.; Kirstein, T.; Locher, I.R.; Mameli, F.; Paradiso, R.; Vozzi, G. Organic field effect transistors for textile applications. IEEE Trans. Inf. Technol. Biomed. **2005**, 9, 319–324.

7. Maccioni, M.; Orgiu, E.; Cosseddu, P.; Locci, S.; Bonfiglio, A. Towards the textile transistor: Assembly and characterization of an organic field effect transistor with a cylindrical geometry. Appl. Phys. Lett. **2006**, 89, 143515:1–143515:3.

8. Cosseddu, P.; Mattana, G.; Orgiu, E.; Bonfiglio, A. Ambipolar organic field-effect transistors on unconventional substrates. Appl. Phys. A **2009**, 95, 49–54.

9. Jang, J.; Nam, S.; Hwang, J.; Park, J.J.; Im, J.; Park, C.E.; Kim, J.M. Photocurable polymer gate dielectrics for cylindrical organic field-

effect transistors with high bending stability. J. Mater. Chem. **2012**, 22, 1054–1060.

10. Toivola, M.; Ferenets, M.; Lund, P.; Harlin, A. Photovoltaic fiber. Thin Solid Films **2009**, 517, 2799–2802.
11. Lee, M.R.; Eckert, R.D.; Forberich, K.; Dennler, G.; Brabec, C.J.; Gaudiana, R.A. Solar power wires based on organic photovoltaic materials. Science **2009**, 324, 232–235.
12. Liu, J.; Namboothiry, M.A.; Carroll, D.L. Fiber-based architectures for organic photovoltaics. Appl. Phys. Lett. **2007**, 90, 063501:1–063501:3.
13. Bedeloglu, A.C.; Demir, A.; Bozkurt, Y.; Sariciftci, N.S. A photovoltaic fiber design for smart textiles. Text. Res. J. **2010**, 80, 1065–1074.
14. O'Connor, B.; Pipe, K.P.; Shtein, M. Fiber based organic photovoltaic devices. Appl. Phys. Lett. **2008**, 92, 193306:1–193306:3.
15. Hamedi, M.; Forchheimer, R.; Inganäs, O. Towards woven logic from organic electronic fibres. Nat. Mater. **2007**, 6, 357–362.
16. Cherenack, K.; Zysset, C.; Kinkeldei, T.; Münzenrieder, N.; Tröster, G. Woven electronic fibers with sensing and display functions for smart textiles. Adv. Mater. **2010**, 22, 5178–5182.
17. Gassmann, A.; Melzer, C.; von Seggern, H. The Li_3PO_4/Al bilayer: An efficient cathode for organic light emitting devices. J. Appl. Phys. **2009**, 105, 084513:1–084513:6.
18. Jonda, C.; Mayer, A.B.R.; Stolz, U.; Elschner, A.; Karbach, A. Surface roughness effects and their influence on the degradation of organic light emitting devices. J. Mater. Sci. **2000**, 35, 5645–5651.
19. Taschuk, M.T.; Hawkeye, M.M.; Brett, M.J. Glancing angle deposition. In Handbook of Deposition Technologies for Films and

Coatings, 3rd ed.; Martin, P.M., Ed.; Elsevier Inc.: New York, NY, USA, 2010; pp. 621–678.

20. Ke, L.; Lai, S.C.; Liu, H.; Peh, C.K.N.; Wang, B.; Teng, J.H. Ultrasmooth silver thin film on PEDOT:PSS nucleation layer for extended surface plasmon propagation. ACS Appl. Mater. Interfaces **2012**, 4, 1247–1253.

21. Farahzadi, A.; Niyamakom, P.; Beigmohamadi, M.; Meyer, N.; Keiper, D.; Heuken, M.; Ghasemi, F.; Rahimi Tabar, M.R.; Michely, T.; Wuttig, M. Stochastic analysis on temperature-dependent roughening of amorphous organic films. Europhys. Lett. **2010**, 90, 10008.

22. Zhang, F.; Xu, Z.; Zhao, D.; Zhao, S.; Jiang, W.; Yuan, G.; Song, D.; Wang, Y.; Xu, X. Influence of evaporation conditions of Alq_3 on the performance of organic light emitting diodes. J. Phys. D Appl. Phys. **2007**, 40, 4485:1–4485:15.

23. Liu, Z.; Song, Y.; Shangguan, Y.; Zheng, Q. Conductive behavior of composites composed of carbon black-filled ethylene-tetrafluoroethylene copolymer. J. Mater. Sci. **2007**, 42, 2903–2906.

24. Lee, S.; Park, J.S.; Lee, T.R. The wettability of fluoropolymer surfaces: Influence of surface dipoles. Langmuir **2008**, 24, 4817–4826.

25. Kolesnikov, V.A.; Zolotarevsky, V.I.; Vannikov, A.V. Anomalous current-voltage characteristics of thin polymer films. Phys. Status Solidi A **2003**, 200, 388–392.

CHAPTER 7

Textile-Based Electronic Components for Energy Applications: Principles, Problems, and Perspective

Vishakha Kaushik, Jaehong Lee, Juree Hong, Seulah Lee, Sanggeun Lee, Jungmok Seo, Chandreswar Mahata and Taeyoon Lee *

Nanobio Device Laboratory, School of Electrical and Electronic Engineering, Yonsei University, 50 Yonsei-ro, Seodaemun-Gu, Seoul 120-749, Korea

ABSTRACT

Textile-based electronic components have gained interest in the fields of science and technology. Recent developments in nanotechnology have enabled the integration of electronic components into textiles while retaining desirable characteristics such as flexibility, strength, and conductivity. Various materials were investigated in detail to obtain current conductive textile technology, and the integration of electronic components into these textiles shows great promise for common everyday applications. The harvest and storage of energy in textile electronics is a challenge that requires further attention in order to enable complete adoption of this technology in practical implementations. This review focuses on the various conductive textiles, their methods of preparation, and textile-based electronic components. We also focus on fabrication and the function of textile-based energy harvesting and storage devices, discuss their fundamental limitations, and suggest new areas of study.

KEYWORDS

wearable electronics; textile electronics; conductive fabric; textile sensors; energy harvesting and storage

1. INTRODUCTION

Electronic devices have the ability to address basic needs [1,2,3,4,5,6]. When electronic devices were first being developed, scientists imagined great leaps forward such as the development of artificial intelligence, and realized the innovation of the Bell telephone in 1876 [7]. A complex number calculator was later developed by the Bell laboratory in 1939 [8]. Seven years after this valuable development, the first fully-functional digital computer, known as ENIAC, came into existence [9]. Although ENIAC was the fastest machine at that time, it was difficult to physically transport due to size ($2.4 \times 0.9 \times 30$ m^3) and power requirements (150 kW). Thus, the primary goals of researchers at this time were to decrease the size and power consumption of electronic devices, which they did with the miniaturization of electronic devices and developed the first personal computer, Kenbak-1, and the first laptop computer, Epson HX-20 [10,11,12]. The increasing pace of technological developments since the late 20th century has resulted in exploration of mobile and wearable devices [13,14]. Early stage wearable computers were composed of backpack systems with uncomfortable wiring networks. Currently, researchers aim to develop flexible wearable devices that integrate seamlessly into regular clothing such as Google glass, the Apple watch, and electronic skins [15,16,17,18]. An illustration of the history of these electronic components from Bell's telephone to smart wearable devices is presented in Figure 1 [7,8,9,10,11,12,13,14,15,16,17,18,19]. Recent market demand has focused on small electronic components for smart wearable device applications, leading to research and products featuring technology that can directly integrate electronic components into textiles or fabrics for a variety of applications including sensors, medical devices, safety instruments, and planar waveguides [4,20,21,22,23,24,25, 26,27, 28,29,30,31].

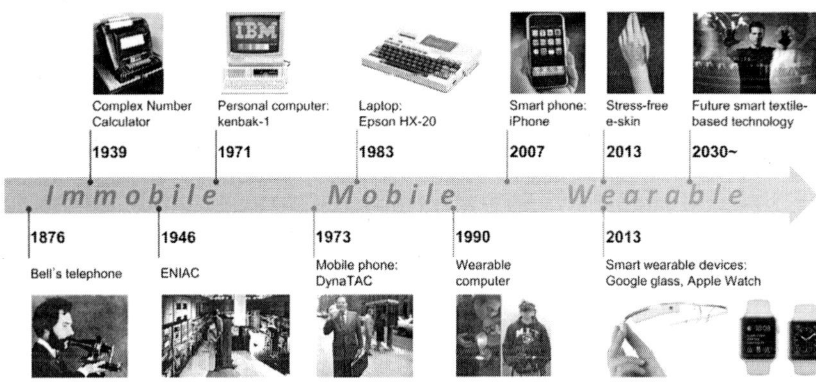

Figure 1. A chronological illustration presenting the journey of electronic components from immobile to mobile, and finally to wearable textile-based electronics.

Textiles have been used by mankind since prehistoric times, and they are essential to daily life. Fashion dictates that the textiles must look attractive and fit comfortably. Traditional textiles are composed of natural or synthetic fibers and filaments. Recent advances in electronics have led to the development of a new generation known as electronic textiles (E-textiles) or smart textiles. E-textiles consist of fabric that provides advanced functions in the form of electronic features or interconnection [32,33]. E-textiles can be categorized into two types:

classical, where the electronic components are embedded into garments and integrated, where the electronic components are directly integrated into textile substrates. E-textiles have attracted attention in technical and socioeconomic fields because of their ability to sense and respond to environmental stimuli [33,34]. Due to this ability, they are ideal components for smart wearable electronics, and represent a new branch of flexible multifunctional fabric electronics. Many research groups have explored the integration of flexible devices such as solar cells, field-effect transistors (FETs), light emitting diodes (OLEDs), and photovoltaic devices into textiles; however, these components cannot operate without sufficient electrical energy, making the development of energy harvesting and storage devices essential [35,36,37,38,39,40,41,42,43,44,45,46].

This paper reviews the development of various key components that are required for E-textiles. The most fundamental element is the conductive textile, which is required for interconnection to transfer power and data. Energy harvesting and storing applications are necessary for powering the electrical components such as transistors, LEDs, and sensors for their utilization in various functional E-textiles. A systematic view is suggested and elucidated with examples.

2. CONDUCTIVE TEXTILE: MATERIALS AND FABRICATION

It is ideal for electronic interconnects to be realized on fibers or fabrics for the fabrication of textile-based wearable devices. The development of highly-conductive textiles for interconnection is crucial to minimize energy loss. Three types of materials have been explored for the fabrication of high-conductivity textiles: polymer-based, carbon-based, and metallic materials. In this section, the features of conductive textiles consisting of organic and inorganic materials will be discussed.

2.1. Polymer-Based Conductive Textiles

Conductive polymers have attracted significant scientific interest for their interesting and unique electrical, optical, and mechanical properties [47,48,49]. They have been explored in a number of applications such as chemical and biological sensors, drug delivery, corrosion protection, nanoelectronic devices, biomedicine, microwave absorption, and electrocatalysis [50,51,52,53,54,55,56,57,58,59]. Shirakawa et al. were the first to demonstrate conductivity in polymers via doping, showing that exposure of polyacetylene (PA) polymer to iodine vapor increases conductivity by up to seven orders of magnitude [60]. The enhanced conductivity of PA is attributed to redox reactions (charge transfer complexes) between the PA and iodine vapor. The search for highly conductive polymers has resulted in the expansion of this area, and a number of conductive polymers have been used in the last 35 years. Various applications such as protective clothing, sportswear, health monitoring, actuators, and sensors for wearable E-textiles have now been explored via these conductive polymer fabrics [61,62,63,64,65].

Conjugated polymer polypyrrole (PPy) has received much attention due to its high conductivity paired with chemical and environmental stability [66,67]. PPy also has the advantage of being easy to synthesize at large areas with different porosities at room temperature. Conductive polymer textiles can be fabricated via polymerization, wet spinning, or dip coating processes [68,69,70]. In situ polymerization or chemical oxidation polymerization are commonly used [71,72]. The experimental steps used to fabricate conductive PPy via in-situ chemical polymerization are illustrated in Figure 2a. Highly-conductive PPy nanofibers with diameters between 60 and 100 nm that are synthesized in p-hydroxyazobenzene sulfonic acid act as a functional dopant. PPy is deposited when the fabric and polymer solutions come into contact due to the liquid-solid interface interaction. This is considered a physical adsorption followed by a polymerization process, and results in a layer of conducting polymer with a smooth surface over the fibers. PPy can also be deposited onto other complex structures. Figure 2b shows the scanning electron microscope (SEM) image of a conductive fiber composed of carbon nanotube bundles coated with PPy [73]; however, brittleness and rigidity are two potential drawbacks to using PPy.

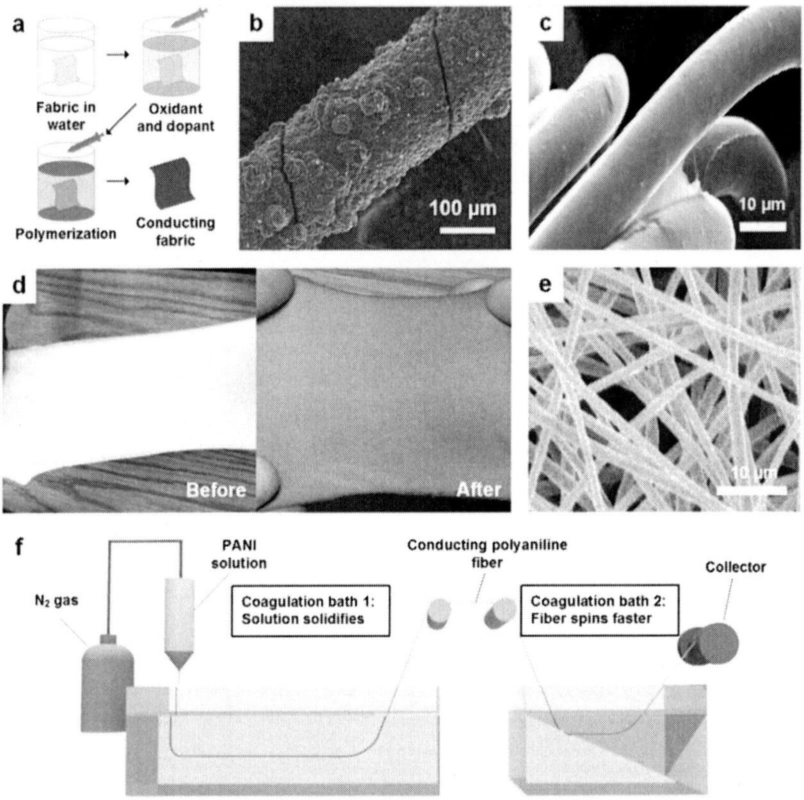

Figure 2. (**a**) Schematic illustration of the in situ chemical polymerization for conducting polymer-coated fabric; (**b**) SEM image of a conductive fiber composed of carbon nanotube bundles coated with polypyrrole; (**c**) SEM micrograph of stretchable spandex fabric after a dip coating process with polystyrene sulfonate-doped poly-(3,4-ethylenedioxythiophene) (PEDOT:PSS); (**d**) Optical image of stretchable fabric before and after dip coating with PEDOT:PSS; (**e**) Surface morphology of polyaniline (PANI): polyethylene oxide (PEO) electrospun fiber; (**f**) Wet fiber spinning technique process. Reproduced from [68,73,74] with the permission by Royal Society of Chemistry, Copyright 2015, and by ACS Publications, Copyright 2010, 2012.

Another interesting conductive polymer is a polythiophene derivative, poly-(3,4-ethylenedioxythiophene) (PEDOT), which shows high electrochemical stability in oxidized form due to its planar structure and delocalized π electrons [75]. Its unique structure contains dioxyalkylene bridging groups at positions 3 and 4 of its heterocycle ring, making it the most valuable among all conducting polymers. Thus, it shows high conductivity and good electrical, thermal, and chemical stability when compared to PPy. Although PEDOT itself is a conductive polymer, it has low solubility, which limits usage. Generally, polystyrene sulfonate (PSS)-doped PEDOT (PEDOT:PSS) is used to obtain a stable dispersion in a water solvent. PEDOT:PSS has been used extensively in the last few years as a conducting polymer [76]. The simplest technique for coating is dip coating, where the fabric is simply dipped into the conductive solution, resulting in a conductive polymer fabric. Ding et al. have reported the preparation of PEDOT:PSS conductive fabrics using the dip coating process [68]. Figure 2c shows the surface morphology of a spandex fabric after soaking in a commercially-available PEDOT:PSS dispersion. It is clear from the Figure 2c that the surface of the fabric is smooth and uniform after conductive coating. The conductivity of the single soaked fabric was reported to be 0.1 S/cm, reaching up to 2 S/cm after multiple soaking steps. The optical image of the PEDOT:PSS-soaked textile is shown in Figure 2d. Thus, conductive fabrics can be produced in a simple manner.

Polyaniline (PANI), known as aniline black, has also been studied as a conductive polymer. PANI revolutionized polymer chemistry due to its many advantages including stability, cost effectiveness, and switching characteristics between conductive and resistive states. Its electrical conductivity is due to the partial oxidation or reduction process and can be tuned to achieve the required conductivity for a given application. It is fabricated via a chemical oxidative polymerization process of aniline, which is a multistep and slightly complex process. Figure 2e shows the surface morphology of electrospun PANI fibers that exhibit high electrical conductivity between 70 and 150 S/cm [74]. However, a more simplified, single-step fabrication method is reported by Pomfret et al. where polyaniline fibers are fabricated using a wet spinning process [77]. The resulting fibers exhibited high conductivity (~1000 S/cm for bulk material) and good mechanical strength (young modulus ~2 GPa). Figure 2f shows the schematic design of the wet spinning process utilized for the production of conductive polyaniline polymer. A mixture of organic solvents and the emeraldine base form of polyaniline is used to make a spinning solution, which then enters the first coagulation bath where it solidifies and forms fibers. In the second coagulation bath, the fibers spin rapidly, resulting in stretching. The stretched fibers are then doped with an acid suitable for the preparation of conducting fibers, as shown in Figure 2f. As discussed earlier, conductive polymers are considered useful in wearable textiles; however, the conductivity values should be modified for their optimization in the production of smart wearable electronics. A comparison of the properties of these conductive fibers is given in Table 1.

Table 1. Summary of materials and their electric properties.

Structure of conductive textile	Material use to prepare or to coat on fibers	Merits of Conductive textiles	Technique used to fabricate conductive textile	Advantages of growth technique	Disadvantages of growth technique	Electrical property	Ref.
Polymer-based	Polypyrrole	Chemical and environment stable	Functional dopant induced process	Novel & Easy process. High Yield easily deposit on fabric surface	Need to optimize dopant concentration for better conductivity results	120–130 S/cm	[78]
	Polyaniline	Cost effective, stable	Wet Spinning	Produced fibers show high electronic and mechanical strength, fabricated thick fibers	Need to combine Individual fibers	140–750 S/cm	[79–81]
	PEDOT:PSS	Highly conductive good thermal and chemical stability	Dip Coating	Simple process	Coating is slow, tide lines can forms	0.4–2.0 S/cm	[68,82]
Carbon-based	Graphene	Highly conductive & stable, high strength	Electrostatic self-assembly with BSA followed by chemical reduction	Easy to attach graphene oxide films to textiles	Multistep process, wrinkles observed on surface of samples	10–20 S/cm	[83,84]
	Carbon nanotubes	Conductive, high mechanical strength	Wet spinning	Continuous long length (cm) CNT yarns	Slow processing	125–3000 S/cm	[85–87]
Metal-based	Al-coated conductive paper	Stable, conductive, stretchable	Chemical solution process	Low cost, Vacuum not required. Direct deposition of Al on fibers	Need a suitable catalyst for decomposition process	19 mΩ/□	[88]
	AgNP/Ag nanowires	Conductive	Wet spinning	Simple, highly conductive material	Need optimization of suitable solvents and coagulation liquids	2200–5400 S/cm	[89,90]

2.2. Carbon-Based Conductive Textile

Carbon-based materials including carbon nanotubes (CNTs), carbon fibers, carbon nanoparticles, and graphene present extraordinary properties such as high mechanical strength, light weight, environmental stability, and superior thermal and electrical conductivity [91,92,93,94,95,96,97,98]. These intriguing properties make carbon-based materials important candidates in wearable electronic textiles. Among all carbon allotropes, the most widely explored are CNTs and graphene. While carbon fibers have been explored in the textile industry for a long time, CNTs are increasingly chosen over commercial carbon fibers because of their lower density, and higher tensile and compressive strength [99,100,101]. CNT can be coated on various fibers such as cellulose or polyester yarns via a simple and cost effective dipping and drying method, as shown in Figure 3a [87]. Here, cotton fibers are simply dipped into a single-walled carbon nanotube (SWCNT) solution, resulting in a black conductive fabric. Figure 3b shows the surface morphology of the SWCNT-coated cotton fibers. The as-prepared CNT-cotton yarns show high electrical conductivity (125 S/cm) with outstanding stretchability and flexibility. The coating of CNTs results in conductivity of fibers and a CNT-cotton thread can emit the light of an LED, as shown in Figure 3c [102].

CNT fibers without any supportive fiber can be made using wet or dry spinning methods. Zhang et al. reported a wet spinning method for the production of long CNTs; a schematic of their wet spinning process is shown in Figure 3d [103]. The electrical conductivity of CNT fibers was enhanced up to ~6.7×10^4 S/cm by iodine-doping [104]. The enhancement in the conductivity of CNT fibers via doping is shown in Figure 3e [104]. Jiang et al. have fabricated CNT yarns using a super-aligned array of CNTs [105]. The way to produce continuous CNT yarns 30 cm in length from a CNT array is similar to drawing a thread from a silkworm cocoon. Figure 3f shows an optical image of the yarn pulled from a CNT array [105]. The authors have constructed a light bulb filament using these CNT yarns and observed that their strength and conductivity is enhanced by heating at high temperatures. The CNT-yarn-based filament emits incandescent light for 3 h, and after this time span the conductivity of the filament increases by up to 13%, while the tensile strength of the CNT yarn increased from 1 mN to 6.4 mN. Thus, these CNT yarns can be woven into fabrics [105]. However, this process has some disadvantages in terms of slow processing and limited to scaling for thick or multistrand forms. Other issues include high junction resistance.

Graphene oxide (GO) has also been widely studied for conductive fibers. Yun et al. presented an electrostatic method for fabricating textiles that are wrapped with reduced graphene oxide (RGO), which makes them conductive and stable (Figure 3g) [83]. Nylon-6 yarns were functionalized with bovine serum albumin molecules to obtain positively-charged surfaces. GO sheets were then electrostatically assembled over the fibers via a simple immersing and shaking process in a GO solution [83]. Finally, the GO are chemically reduced to RGO. They have also demonstrated the lightning of an LED using the synthesized conductive fibers, a process that works well even after folding, as shown in Figure 3h [83]. Freestanding RGO fibers can also be synthesized using wet or dry spinning methods. Xu et al. have reported porous graphene oxide with good electrical conductivity ($\sim 4.9 \times 10^3$ S/m) and high specific surface area (884 $m^2 \cdot g^{-1}$) [106]. This graphene-based hybrid structure presents great potential in fiber-based electrochemical super-capacitors [84]. The above described properties of carbon-based materials open a new path in electronic textiles for wearable electronics. However, further investigation is needed to look into the possible side effects of wearable electronics, such as toxicity with skin contact and stability in air.

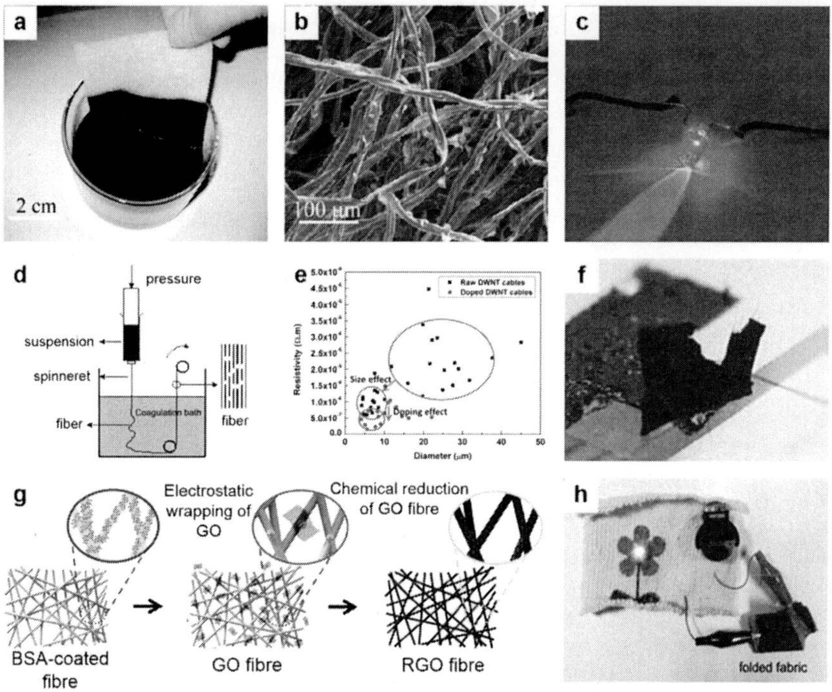

Figure 3. (**a**) Optical image shows the dip-coating process of fabric into single-walled carbon nanotube (SWCNT)s ink for SWCNT-based conductive fabric; (**b**) SEM image of a cotton sheet coated with SWNTs on the fibrous surface; (**c**) Demonstration of LED emission with CNT-based conductive fibers; (**d**) Schematic of the experimental setup of the wet spinning method for CNT-based conductive fibers; (**e**) Resistivity versus diameter plot for raw DWNT (double walled CNT) and iodine doped DWNT cables; (**f**) Image shows a free standing CNT array through which a yarn is pulling out; (**g**) Schematic illustration of the fabrication process of graphene oxide-based conductive fiber using electrostatic force; (**h**) Photograph of a LED light using a folded, reduced graphene-oxide-based conductive fabric. Reprinted from [83,87,102,103,104,105] with the permission by ACS Publications, Copyright 2010, 2008 and by John Wiley and Sons, Copyright 2008, 2013 and by Nature Publishing Group, Copyright 2002, 2011.

2.3. Metal-Based Conductive Textiles

Metallic materials are considered a better choice for wearable electronics because of their high conductivity [89,90]. Vacuum deposition is one of the most widely used techniques for metal deposition and is also frequently used for microfabrication processes. Sputtering or evaporation methods can be used to directly form a thin metal layer on the exposed surface of the fibers [107,108]; however, these techniques have huge drawbacks such as expensive instruments, limited sample size, and non-compatibility for batch process. Also, these methods are not suitable for depositing metal on complex structures such as porous fibers, and the deposited films are weak against repeated deformations like folding or bending. In order to avoid these issues, alternative methods must be used to synthesize metal-based conductive textiles.

Electroplated or electroless plated metals, liquid metals, and metallic nanomaterials such as nanoparticles or nanowires are considered ideal candidates in the growing field of wearable electronics [109,110]. Metal plating methods have been widely used to deposit metal for printed circuit boards, metallization processes in microfabrication, and corrosion protection. Figure 4a shows an SEM image of a gold-plated conductive fiber, in which Kevlar fibers were first treated with Sn^{2+} and Pd^{2+} ions followed by electroless Ni deposition for conductivity [109]. Au was then electroplated onto the surface to achieve a conductive fiber with a conductivity of ~6 S/cm and power output of 1 W. However, the metal films are weakly attached to the surface and may separate from the fiber during further processing. To overcome these obstacles, a recent study presents a new, simple, and inexpensive chemical solution method, as shown in Figure 4b [88]. In this process, an Al precursor was used to deposit Al on the surface of fibrous materials. Figure 4c shows the electrical resistances verses immersion time in the Al precursor solution, and the inset shows an LED connected to these fibers [88].

Another method of synthesizing metal-based conductive textiles is the use of liquid metals [111]. Eutectic gallium indium (EGaIn) is a liquid metal with high conductivity ($\sigma = 3.4 \times 10^4$ S/cm) and a low melting point ($T_m = 15.7$ °C) that spontaneously forms a thin oxide layer that can be used to mold the metal into complex shapes [111]. When sufficient force is applied, the oxide layer can be damaged and the shape of the EGaIn can be changed. Zhu et al. has filled a hollow elastomer fiber with EGaIn to obtain stretchable conductors, as shown in Figure 4d [112]. The authors observed that the metallic continuity is maintained even after the applying a 700% strain.

Various metallic nanomaterials such as silver, copper, aluminum, gold, and tin are used as conductive materials for textiles [113]. Atwa et al. demonstrated nylon, polyester, and cotton threads dip-coated with Ag nanowires (AgNWs), which exhibit high mechanical flexibility and a small increase in resistance after 200 bending cycles as compared to commercial conductive fibers [113]; however, metallic nanomaterials mixed with polymers can lead to better stability. Recently, Lee et al. reported the fabrication of silver nanoparticle (AgNP)/polymer-coated Kevlar fiber with a silver content of more than 80 wt % [20]. Figure 4e shows the surface morphology of the as-prepared fibers along with an inset showing the AgNPs on the surface [20]. In this work, the $AgCF_3COO$ solution was absorbed by a poly (styrene-block-butadienstyrene) (SBS) polymer and then reduced to AgNPs using hydrazine hydrate. Figure 4f shows the small change in resistance during 3000 folding tests [20]. In another study, Lee et al. inserted AgNWs into a stretchable AgNP/SBS structure to obtain increased conductance under strain [90]. The fiber shows high electrical conductivity, 2450 S/cm, and can stretch up to 900% strain as shown in Figure 4g [90]. Table 1 presents a comparison of the properties of these metallic conductive fibers.

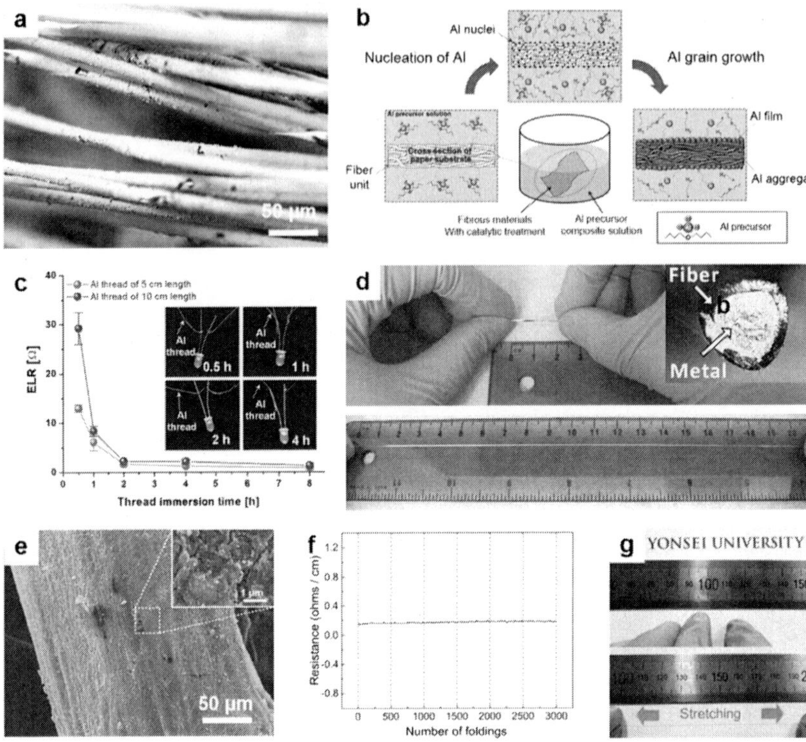

Figure 4. (**a**) SEM image of conductive Kevlar fibers coated with Ni/Au; (**b**) Schematic representation of the chemical solution process for the fabrication of conductive Al textiles; (**c**) Plot of the electrical linear resistance of Al-based conductive fiber synthesized by the chemical solution process and inset shows the functioning LEDs; (**d**) Optical image of unstretched and stretched (up to 20 cm) conductive fiber; (**e**) SEM image showing the surface of the Ag nanoparticles and elastomer composite-based conductive fiber; (**f**) Electrical resistance change in the Ag-based conductive fiber during the 3000 cyclic folding test; (**g**) Optical image of pre- and 900%-strained Ag nanoparticles and elastomer composite fibers. Reproduced from [20,88,90,109,112] with the permission by ACS Publications, Copyright 2011 and by John Wiley and Sons, Copyright 2013, 2013, 2015, 2015.

3. TEXTILE ELECTRONIC COMPONENTS

Textile-based electronic devices require electronic functionalities on fiber/textile materials with robust mechanical properties. In this section, we will discuss electronic devices based on fibers and their use in textile electronics. The operation of electronic devices and their stability in a textile system will be explored in detail. Specifically, we will demonstrate recent developments in textile-based electronic devices such as transistors, light emitting diodes (LEDs), and sensors.

3.1. Transistor

The integration of electronic devices into textiles at the fiber level is vital to the future of wearable electronics. There are various aspects to be considered before employing electronic devices into a textile-based system such as materials, structures, and the compatibility of electronic devices with textile fabrics. As a fundamental building block of electronic devices, transistors must be integrated into fabric or textiles. In this section, we briefly introduce fiber or textile-based transistors and their operation. In general, an electronic device that can amplify and convert electronic signal to electrical power is known as transistor. Since transistors are basic components in electronic devices, the integration of transistors into fabric or textile systems is highly important. However, conventional inorganic FETs cannot be applied to fabric or textile-based electronics due to flexibility issues and limitations in the fabrication process. Given the technological limitations of inorganic materials, organic thin film transistors (OTFTs) have received significant attention owing to their flexibility, compatibility with flexible electronics, light weight, biocompatibility, low cost, and low-temperature process capability [114,115,116,117].

Organic-based transistors due to their flexibility, easy fabrication process and comparable electrical performance (as compared to inorganic TFTs) are desirable in textile industry [118,119,120]. Two types of fiber-based

OTFTs have been reported; fiber organic field effect transistors (F-OFETs) and organic electrochemical transistors (OECTs) [121]. The working principle of F-OFETs is similar to that of commercial FETs. In O-FETs, organic thin-film semiconductors are used as an active layer instead of inorganic materials. Figure 5a shows a schematic diagram of F-OFETs. Here, a commercially available cylindrical metal fiber (Al & stainless steel) is insulated using a uniform layer of polyimide, followed by the evaporation of pentacene as an active material PEDOT:PSS is then deposited on the pentacene surface using a soft lithographic process to form source and drain electrodes. Figure 5b,c shows an optical microscope image of the channel length (~45 μm) and corresponding I_D-V_D curve of an F-OFET [122]. In spite of the low value of the width/length ratio of the F-OFET, the on current and on-off ratio are reasonably high (~10^4), which is fully comparable with those of planar devices. Although F-OFETs exhibit decent electrical properties, they suffer from significant disadvantages such as high operating voltages, high equipment cost, and a complicated fabrication process. OECTs have been considered as an alternative to F-OFETs, since they can be fabricated on the basis of fiber structure. The working principle of OECTs is based on the electro-chemical process, where a reversible process of doping and de-doping of electronic polymers allows the transistor to operate. Figure 5d shows a schematic picture of OECTs on a planar substrate [123]. Here, source (S), drain (D), and gate (G) electrodes are demonstrated on flat substrates such as glass, plastic, and paper. The conducting polymer PEDOT/PSS films are first micro-patterned on a flat substrate, and are followed by the pattering of an electrolyte (micro-channel of ~3 μm height and ~15 μm width) over the patterned PEDOT/PSS film.

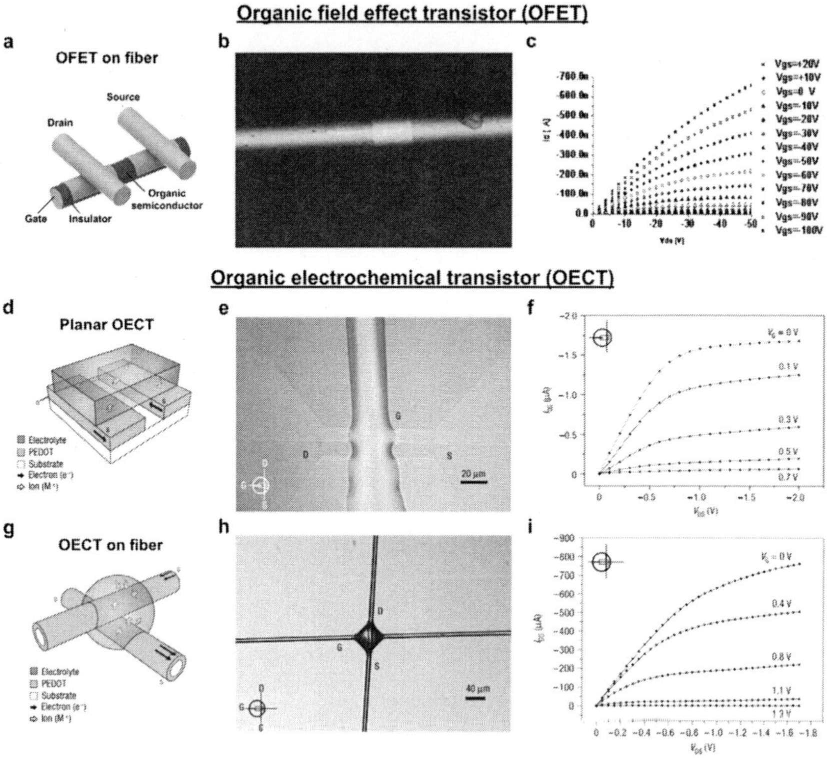

Figure 5. (**a**) Schematic picture of the organic field effect transistor (O-FET); (**b**) Optical image showing channel area; (**c**) The output characteristic of O-FET; (**d**) Schematic picture of a planar OECT; (**e**) Optical micrograph of a planar organic electrochemical transistor (OECT); (**f**) Output characteristic of a planar OECT; (**g**) Schematic picture of a fiber OECT; (**h**) Optical micrograph of a fiber OECT; (**i**) Output characteristic of a fiber OECT. Reprinted from [122,123] with permissions by AIP Publishing LLC, Copyright 2006 and by Nature Publishing Group, Copyright 2007.

An optical micrograph of this configuration is shown in Figure 5e, and the corresponding output characteristic of a planar ECT is shown in Figure 5f [123]. The output characteristics clearly indicate that the current is saturated with increasing drain voltage, and that the gate voltage is low compared to that of F-OFETs. When the gate voltage is zero, the transistor is in the ON state, while the gate voltage transistor moves to the OFF state when the gate voltage is further increased, as shown in Figure 5f. In order to fabricate OECTs on a fiber structure, a simple new method was performed by Hamedi et al. [123]. Figure 5g shows a schematic picture of a fiber OECT constructed using PEDOT-coated polyamide fibers (~10 μm diameter). Two PEDOT-coated fibers are kept in a cross geometry and an electrolyte is dropped at the fiber junction, as shown in the optical image in Figure 5h [123]. In this cross geometry, one PEDOT-coated fiber acts as source and drain and the other as gate electrode. An active material (electrolyte with 33 wt % PSS, 12 wt % glycol, 8 wt % sorbitol, water and 0.1 M $NaClO_4$) acts as a conducting channel, which is controlled by the gate electrode. In this way, micrometer-sized fiber OECTs can be easily integrated into textiles. The output characteristics of fiber OECTs are similar to planer OECTs, as shown in Figure 5i [123]. The similar electrical characteristics between fiber OECTs and planer OECTs are attributed to the fact that the operation of electrochemical transistor is dominated by the interface effect between the conducting polymer and electrolyte. Therefore, fiber OECTs can be used as effective transistors in spite of completely different geometry when compared to planar OECTs. Theoretically, the fiber OECTs could be used for large scale production: ~100,000 transistors/cm^2 can be constructed using 10 μm fibers. Also, the authors have fabricated a binary tree multiplexer structure by using these fiber OECTs. This multiplexer is able to encode the information from a large number of data sources into a single channel. Hamedi et al. fabricated a fiber-based OECT using poly 3-hexylthiophene (P3HT) with 100 μm channel length and showed that the fiber can be operated below ~1 V [124].

Since fiber OECTs have many advantages such as low operating voltages, large amounts of current densities, and low cost production, they can easily replace OFETs and can be integrated to wearable electronics by directly weaving them into fabrics. Bonderover et al. have demonstrated textile inverters made by weaving fiber-based transistors [125]. They have shown that E-textiles with electronic components can be designed by simply woven fibers. Thus, in the future it is possible to fabricate such E-textiles. However, their long response time and low switching frequency remain significant hurdles. In addition, the exact charge transport mechanism of fabric OECTs must be revealed to fully understand the operation process.

3.2. Light-Emitting Diodes

A light emitting diode (LED) is a semiconductor device that emits visible light when electron hole recombination takes places inside the device. Since the first commercial infrared inorganic LED was presented by Holonyak et al. in 1962 [126], various LEDs have been developed around the world such as visible, ultraviolet, and infrared LEDs [127,128,129,130,131]. Since a solid semiconductor material is generally used to generate light in LEDs, LEDs were originally considered solid-state devices. A breakthrough occurred in 1987 when Tang and Van Slyke reported a new type of LED based on organic semiconductors [132]. Organic LEDs (OLEDs) have several advantages, including lower manufacturing cost, faster response time, potential flexibility, low power consumption, and less heat dissipation [133]. Since OLEDs are malleable, they can be compatible with flexible substrates and can be applied to wearable electronics. Current research into OLEDs focuses primarily on two-dimensional (2D) devices, however, one-dimensional (1D) OLEDs have begun to receive attention for future wearable electronics. In this section, various 1D fiber-based OLEDs will be discussed.

Connor et al. fabricated a fiber-based OLED using molecular organic compound as an active layer [134]. The inherently-flexible nature of molecular organic compounds is compatible with fiber structure. Figure 6a shows a schematic illustration of the fiber-based OLED structure used in this study [134]. Here, a polyimide-coated silica fiber was prepared as a substrate, and emission layers and organic charge transport layers were deposited between the anode and cathode electrodes via a vacuum thermal evaporation technique. A photograph of a 1 mm segment of green-light OLED on fiber clearly shows the emitting characteristics of green light. For a performance comparison of electricity and luminescence between a fiber-based OLED and planar OLED, they fabricated a planar OLED using a planar silicon/polyimide substrate and investigated the electrical and luminescence characteristics. The electrical and luminescence characteristics of fiber-based OLEDs showed comparable results to planar OLEDs; however, a slightly larger leakage current was measured at low bias due to the surface roughness of the fiber structure. In addition, the fiber is rather thick at ~480 μm, which can hinder practical use. Yang et al. recently fabricated single core-shell fiber OLEDs based on an ionic transition-metal complex (iTMC) using a coaxial electrospinning process [135]. A galinstan liquid metal core and a ruthenium tris (bipyridine) with poly (ethyl-oxide) (PEO) mixture shell were co-electrospun to from the cathode and electroluminescent layer, respectively. The anode was formed via thermal evaporation of indium tin oxide. Figure 6b shows a schematic illustration and optical image of an emitting iTMC [135]. The device was initially turned on at 4.2 V and bright electroluminescent was observed at 6.8 V. This work shows the possibility of direct fabrication of 1D flexible OLEDs without flexible substrates; however, the efficiency or brightness of the devices is not clearly shown.

Vohra et al. fabricated luminescent F8BT/PEO nanofibers using electrospinning [43]. Figure 6c shows a schematic illustration and fluorescence microscopy image of the conjugated polymer electrospun nanofiber-based OLED [43]. To fabricate the device, a thin layer of PEDOT-PSS was spin coated onto the ITO/glass substrates, and the F8BT/PEO nanofibers were directly collected onto the resulting layer. Then, a PVK (poly N-vinylcarbazole) solution was spin coated on top, and finally a thin layer of calcium and aluminum were evaporated. Here, the PVK-coated F8BT/PEO nanofibers functioned as an active layer. This work solved the charge injection issue in organic nanofibers by annealing the device at 150 °C, which changed the active layer into ribbon-like morphology. The measured external quantum efficiencies and luminance were 0.5% and 2300 cd·m^{-2} at 6 V, respectively.

Most recently, Zhang et al. developed a color-tunable and wearable fiber-shaped polymer light emitting electrochemical cell (PLEC), which allows for the integration of wearable LEDs in textiles [42]. The fiber-shaped PLEC was fabricated using an all-solution processing method, which increases the chance for practical use. Figure 6d presents a schematic diagram of a flexible fiber-based PLEC [42]. To fabricate the fiber-shaped PLEC, a stainless steel wire was dip coated with ZnO nanoparticles to form an electron transfer layer and decrease leakage current. Then, an electroluminescent polymer layer (a blend of PF-B, ethoxylated trimethylopropane triacrylate and lithium trifluoromethane sulphonate) was grafted over the ZnO-coated stainless steel wire, followed by wrapping the modified steel wire with the aligned carbon nanotube (CNT) sheet. The aligned CNT provided high electrical conductivity and flexibility in the device. Since the fiber-shaped PLEC is a 1D structure, the brightness was the same in all directions. In addition, the authors successfully integrated PLECs into fabrics and showed the tunable colors. This work shows promising benefits such as flexibility, light weight, high brightness that is maintained after bending, and tunable colors, as well as the possibility of integration into wearable electronic devices.

3.3. High Sensitivity Sensors

Sensors are widely used electronic devices which detect specific the characteristics of their surroundings and are present in a wide variety of applications such as electronics, aerospace, medicine, robotics, and machinery [136,137,138,139,140,141,142]. Types of sensors include temperature, pressure, humidity, gravity, sound, and vibration sensors [143,144,145,146,147,148,149,150]. In this review, we focus on textile-based pressure sensors, as sensing pressure is one of the most important functions of wearable electronics.

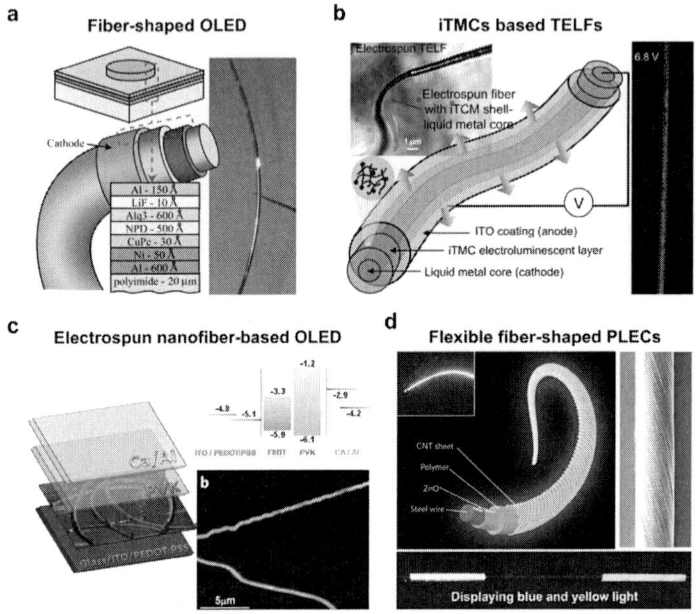

Figure 6. (**a**) Schematic illustration of a fiber-shaped organic light emitting diodes (OLED) structure compared with a typical OLED structure and a photograph of a flexed fiber with a green light-emitting pixel; (**b**) Schematic illustration and optical image of ionic transition-metal complex (iTMCs)-based electro-luminescent nanofibers (TELFs) fabricated via coelectrospinning and luminescence response of TELF under an applied voltage of 6.8 V; (**c**) Schematic of the conjugated polymer electrospun nanofiber-based OLED structure (left), energy level diagram (right top), and fluorescence microscopy image of an electrospun nanofiber, F8BT-PEO (right bottom); (**d**) Schematic illustration showing the structure of flexible fiber-shaped polymer light-emitting electrochemical cells (PLECs). Left inset: photograph of a fiber-shaped PLEC biased at 10 V, right inset: aligned CNT sheet wrapped around the modified stainless steel wire, and bottom inset: PLEC displaying blue and yellow light. Reproduced from [42,43,134,135] with permissions by John Wiley & Sons, Copyright 2007 and by ACS Publications, Copyright 2012, 2011 and by Nature Publishing group, Copyright 2015.

In 1996, a full-body, soft, flexible sensing suit was developed for a small robot by Japanese researchers [151]. They used an electrically-conductive fabric sheet and conductive strings with a spacer, and measured the contact resistance after applying pressure, demonstrating a promising remote robot sensor capable of controlling robotic body interactions. Shimojo et al. fabricated a tactile sensor using pressure-conductive rubber [152] via the development of a thin, flexible, attachable, tactile sensor that can be operated on curved surfaces. They measured the change in resistance under constant strain, observing a change in resistance during a period of 100 s. Their tactile sensor, when properly operated on a robot hand, showed good durability under repeated bending tests. Takamatsu et al. developed a large-area pressure sensor on a meter-scale using dye-coated conductive polymer (PEDOT:PSS) and perfluoropolymer dielectric film [153]. The sensor was constructed by weaving fibers into an area of 16 × 16 cm^2. The sensitivity of this design ranged from 0.98 to 9.8 N/cm^2, sensitive enough to detect human touch and therefore suitable for wearable keyboards and healthcare systems.

Recently, Lee et al. fabricated a textile-based pressure sensor using Ag-SBS composite conductive fibers [20]. A capacitive type of textile-based sensor was fabricated using a poly(dimethyl-siloxane) (PDMS) coating on the conductive fiber and cross-stacking two PDMS-coated fibers, as shown in Figure 7a [20]. Figure 7b shows the relative changes in capacitance of fiber-based pressure sensors via increasing loads of applied pressure [20]. In this operation, the conductive fiber-based pressure sensor showed stable responses to the various loads (Figure 7c), and detected the loading and unloading of grains of extremely low weight. This work further demonstrates e-textile applications with conductive fiber-based pressure sensors. The textile-based pressure sensor was pixelated through weaving techniques and detected the spatial distribution of external pressures, as shown in Figure 7d [20]. This textile-based pressure sensor was also sewn into a smart glove and clothes and showed the control of drone and spider robots, as shown in Figure 7e

[20]. This work demonstrates that human-machine interfaces can be successfully operated using wearable textile-based pressure sensors with ultrahigh sensitivity and stability.

Ma et al. presented a new means of fabricating stretchable fabric [154]. To make stretchable knitted fabric, conductive stretchable fibers were first fabricated via wet-spinning. Figure 8a shows the surface morphology of conductive stretchable fibers composed of Ag nanoparticles, nAg-MWCNTs, and poly (vinylidene fluoride-co-hexafluoropropylene) (PVDF-HFP) matrix [154]. Stretchable fabric was then made by knitting the conductive stretchable fibers. The knitted fabric showed reversible characteristics under 100% tensile strain with mechanical and electrical stability (Figure 8b) [154]. In addition, the stretchable fabric was applied to a robotic finger system and demonstrated feasibility for practical application.

More recently, Lee et al. fabricated highly-stretchable conductive fibers composed of Ag nanowires (Ag NWs), Ag nanoparticles (Ag NPs), and SBS polymer using the wet spinning method [90]. Ag NWs allow for significant stretching capability in the fabric because they are aligned along the applied uniaxial strain and bridge the disconnected networks of Ag NPs inside the fiber, as shown in Figure 8c [90].

Figure 8d shows changes in conductivities with regard to various applied tensile strains [90]. As tensile strain increased, conductivities initially decreased, but improved again with increased strain. The reliability test showed stable operation under a repeatable bending test (Figure 8e) [90]. The authors further demonstrated the capability of detecting human motion using the fabricated stretchable sensor. The smart glove was integrated with fibers on each finger and successfully detected finger sign language. Figure 8f shows the detection of the letters “Y” and “S” via the bending of fingers [90].

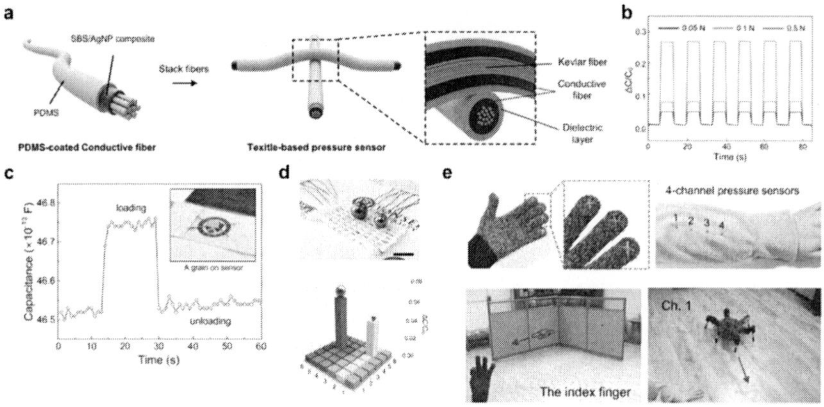

Figure 7. (**a**) Schematic structure of a textile-based pressure sensor fabricated with PDMS-coated conductive fibers; (**b**) Capacitive response of the pressure sensor for the various applied loads; (**c**) Capacitive response of the pressure sensor to the placing and removal of a grain (8 mg); (**d**) Fabrication of pixelated pressure sensor array; (**e**) Photographs of the smart glove and clothes with textile-based pressure sensors and operation using smart glove and clothes to control the drone and spider robot. Reprinted from [20] with permissions by John Wiley & Sons, Copyright 2015.

4. TEXTILE-BASED ENERGY APPLICATIONS: ENERGY HARVESTING & STORAGE

Although commercial energy harvesting and storage devices using semiconducting thin film exhibit high performance, they also have limitations including their complex fabrication process and high cost. Recent novel materials investigated for energy related devices include dye, organic and inorganic polymers with high efficiency, large capacity, abundant material sources, easy fabrication, and flexibility. Along with the development of materials for energy devices, there are several studies on novel methods for energy harvesting using the human body, such as piezoelectric, triboelectric, and thermoelectric nanogenerators. Nanogenerators can provide self-powered systems that are useful in wireless devices. In particular, fiber-based nanogenerators have been primarily studied in the form of twisted, conjugated, or weaved multiple fiber electrodes integrated with supercapacitors.

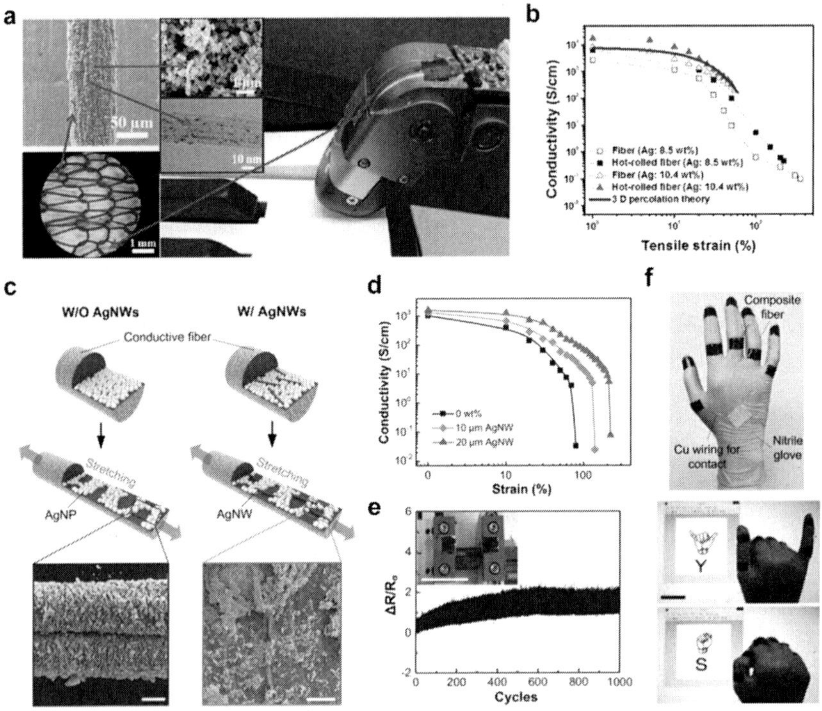

Figure 8. (**a**) FE-SEM and HRTEM images of knitted fabrics made from highly-conductive stretchable fibers. Bottom of this image shows an optical image of 2 ply rope. The knitted fabric was attached to a robot finger (as shown by arrow) for touch sensing; (**b**) Plot shows the conductivities of fibers as a function of tensile strain; (**c**) Schematic illustration of silver nanowires (AgNWs) and silver nanoparticles (AgNPs) in the composite fiber. (Left bottom image in navy colour line: SEM image of AgNP-mixed styrene-block-butadienstyrene (SBS) fiber without AgNWs at 50% strain and and Right bottom image in red colour line: SEM images of the AgNW-AgNP embedded SBS fiber at 50% strain); (**d**) Changes in conductivity of the composite fiber depending on

the length of the AgNWs; (**e**) Reliability test measured the changes in the normalized resistance of the composite fiber; (**f**) Photograph of a smart glove made with composite fibers and motion detection of English letters using a sign language. Reproduced from [90,154] with permissions by ACS Publications, Copyright 2014 and by John Wiley & Sons, Copyright 2015.

4.1. Energy Harvesting from Human Motion

As the number of applications related to mobile electronic devices increases drastically, the development of long-lasting power sources for these devices is increasingly in demand. Sustainable self-powered sources have received significant attention as alternatives to conventional rechargeable batteries. Recently, energy harvesting from human body motions has proven a new means of operating many mobile electronic devices without the need for external power sources [155].

The piezoelectric effect is a unique ability of materials with polarization domains or non-centrosymmetric structures. These materials can generate an electrical charge in response to mechanical stresses such as compression, twisting, or distortion. One novel approach using the piezoelectric effect converts human mechanical motion into electrical energy. Wang et al. demonstrated that ultrasonic waves were converted into electricity using a Pt-coated zigzag electrode with vertically-aligned ZnO nanowires (NWs) [156]. They optimized the size and shape of the ZnO NWs and obtained high-output power per unit of area reaching ~10 μW/cm^2. The generated energy can potentially facilitate long-term sensors that are applicable to mobile electronic devices and systems [157]. Zhang et al. reported a flexible nanogenerator (FNG) fabricated using $BaTiO_3$ nanowire and polyvinyl chloride (PVC) composite piezoelectric fiber [158]. After optimizing their process parameters, they placed FNG on a bent human arm, producing voltage and current outputs up to 1.9 V and 24 nA, respectively. Several methods have been developed for harvesting electrical energy by converting mechanical energy into electricity using piezoelectric materials [159]. A prototype electrostatic non-resonant generator has been designed and optimized by Mitcheson et al. [160]. Using flexible polymer membranes, they can generate an output voltage of 250 V when the generator is pre-charged to 30 V. Qin et al. have described the concept of textile-based piezoelectric nanogenerators that can harvest electrical energy from the friction between two fibers [161]. Although only a small amount of output current (~±5 pA) was obtained during the pulling/releasing cycles of the nanogenerator, they established a novel method of energy harvesting using fabrics.

Further, Jeong et al. reported a novel approach to fabricating a high-performance hyper-stretchable nanogenerator (SEG) with extremely long Ag NWs [162]. The hyper-stretchable piezoelectric elastic composite (PEC) rubber polymer was composed of very long nanowire percolation (VLNP) electrodes and a well-dispersed mixture of lead magnesioniobate lead titanate (PMN-PT) particles and multi-walled carbon nanotubes (MWCNTs). The composite exhibited high stretchability over 200% strain without any mechanical cracking; this could be attributed to variation in the piezoelectric dipoles in the PEC caused by stretching, and helps accumulate electrons, generating voltage and current. An open-circuit voltage (~4 V) and short-circuit current (~500 nA) were obtained and the generated energy can sufficiently operate commercial electronic units. The SEG was tested in biological motion after being sewn onto the knee of a stretchable stocking, as shown in Figure 9a,b [162]. However, textile-type energy harvesters seem to face some prevalent problems such as low energy conversion efficiency, wide spectra of vibration frequencies, and time-dependent amplitudes.

Recently, triboelectric nanogenerators (TENG) have been developed that can harvest electrical energy from mechanical friction, exhibiting high performance, a simple fabrication process, cost-effectiveness and green technology [163,164,165,166]. The principle of TENG operation is the combination of triboelectric and electrostatic induction [166,167]. When two different polymer surfaces are rubbed or compressed, opposite sign electrostatic charges are generated on each surface of the polymer films. Due to the formation of a dipole layer, triboelectric potential in TENG is generated and drives electrical charges through the external load, resulting in the generation of an open circuit voltage and a short circuit current. Lee et al. reported a textile, substrate-based triboelectric nanogenerator (T-TENG) fabricated using an assembly of Au-coated fabric and polydimethylsiloxane (PDMS) [165]. To enhance the surface friction energy, thermally-evaporated Al nanoparticles (NPs) were conjugated with Au-coated textile top electrodes. Triboelectric energy was generated when two textile electrodes were repeatedly compressed and released. The validation of T-TENG in energy harvesting using human motion was further examined on a human arm. To utilize this device in practice, they demonstrated that the alternating current (ac) output signals obtained from the T-TENG were converted to direct current (dc) using a bridge rectifier. Using the rectified dc current, a commercial capacitor was charged and a light-emitting diode (LED) was turned on. When the T-TENG attached to the human arm was bent and released, the generated output voltage and current were 139 V and 39 μA, respectively. The high efficiency can be attributed to enhanced surface roughness induced by uniformly-distributed Al NPs on the textile electrodes.

The abovementioned nanogenerators were produced on flexible, metallic-based electrodes. However, to realize the true potential of wearable electronics, it is necessary to study textile-based flexible energy harvesters without metal electrodes. Metal-free, fiber-based TENG capable of converting biomechanical motion into electricity were introduced by Zhong et al. [163]. The TENG were produced using two kinds of pretreated cotton threads, each coated with multi-walled carbon nanotubes (CNT) and polytetrafluoroethylene-carbon nanotubes. These two fibers were entangled with each other to form a light-weight and flexible fiber-based generator (FBG). When the entangled fibers were stretched, the inter-fiber gap was reduced and the CNT layer on the CNT-coated cotton thread (CCT) became positively charged due to induction, creating free electrons. Woven FBG was applied to a "power shirt" with a wireless temperature sensor capable of charging to a capacity of 10 nF, as shown in Figure 9c [163]. Figure 9d shows that the rectified DC voltage obtained from the FBG was 2.4 V with the charging time of 27 s, showing that the FBG is capable of lighting a red LED. Since the fabricated power shirt can monitor body temperature cost-effectively due to the self-powered TENG system, it makes viable the possibility of woven textiles in the field of mobile medical systems.

Recently, interest in thermoelectric materials has increased due to their ability to generate energy from waste heat. In the presence of temperature gradients, charges in the thermoelectric material are diffused from the hotter to the colder side, inducing electrochemical potential inside the material. In wearable electronics, fiber-based thermoelectric generators (TEGs) are a unique energy harvesting device converting body heat into a reliable power source [168,169,170]. Since human body heat is a steady, plentiful, and convenient source of heat, it is meaningful and useful that TEG can harvest waste body heat. Kim et al. described a wearable TEG utilizing a combination of fabric and printed thermocouples that can be easily embedded in clothing to harvest low-temperature energy [169]. To prevent skin irritation, polymer-based fabric was used as a substrate and a conductive silver thread was used for electrical connection. When the TEG was integrated in a shirt, the temperature of the skin was maintained at 32 °C under cold (5 °C) or hot (25 °C) environments. Interestingly, the output power under cold environment was significantly higher at ~147 nW than that of the hot ambient environment, at ~8.1 nW.

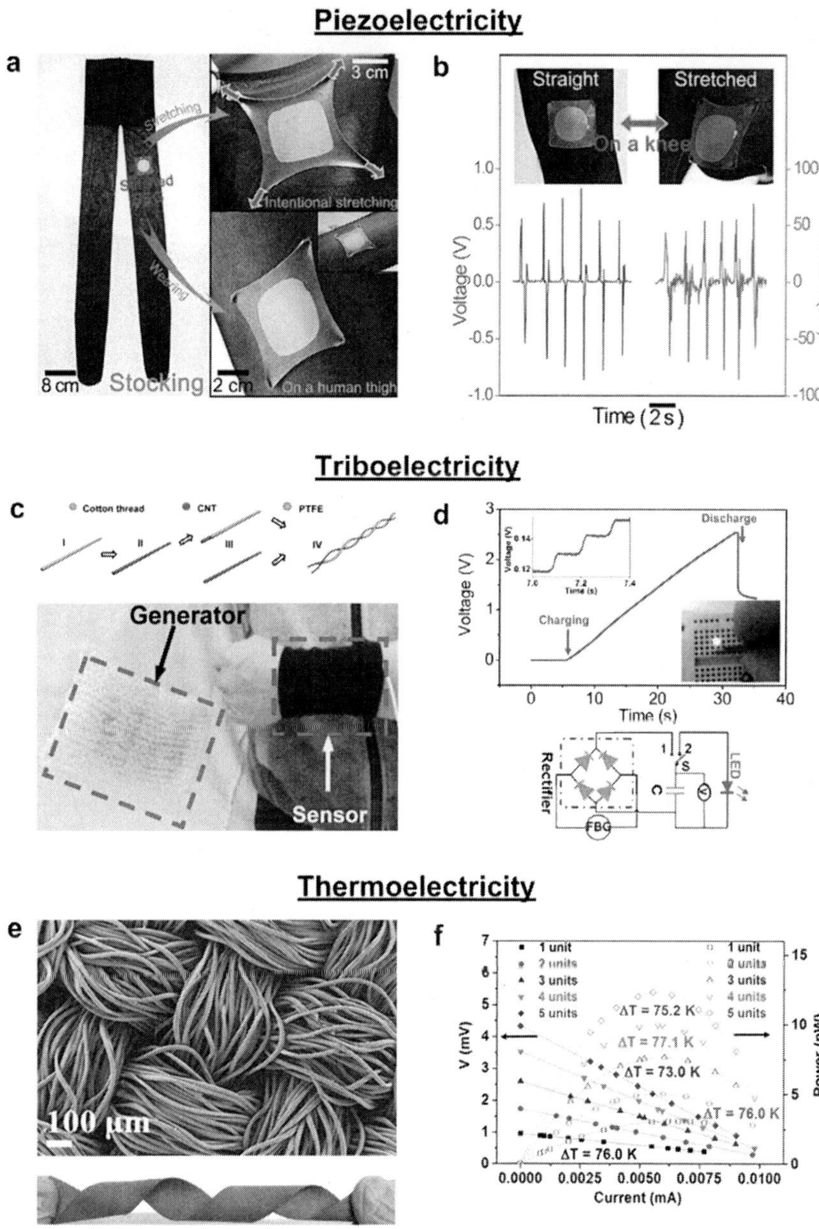

Figure 9. (**a**) Digital photographs of the stretchable elastic-composite generator (SEG) stitched on a nylon stocking. Photograph of the SEG intentionally stretched and fitted over a human thigh; (**b**) Stretching on human knees and the generated voltage and current from the SEG on the

stocking by bending and straightening the knee; (**c**) Schematic diagram illustrating the detailed fabrication process of a fiber-based generator (**above**) and woven fabric FBG sewed on a lab coat (**below**); (**d**) The voltage charging curve of the 2.2 μF commercial capacitor by the "power shirt" and typical bridge rectifier circuit for transforming alternating current to direct current; (**e**) SEM image (**above**) and digital photography (**below**) of the polyester fabric after PEDOT:PSS coating; (**f**) Experimental results of output voltage and power as a function of current for the fabricated devices. Current–power curves are parabolic for all devices. Reproduced from [162,163,168] with permissions by John Wiley & Sons, Copyright 2015 and by ACS Publications, Copyright 2014 and by Nature Publishing Group, Copyright 2014.

To investigate the suitability of thermoelectricity in wearable power generators, Du et al. reported a textile-based TEG featuring a long operating time, easy maintenance, and high reliability [168]. This TEG overcame the common problems of toxicity, poor processing ability, and impermeability by introducing a PEDOT:PSS conducting polymer. The inset of Figure 9e shows the morphology of polyester fabric coated with PEDOT:PSS, which had excellent flexibility in any shape [168]. The PEDOT:PSS-coated fabrics were connected using silver wires, as shown in Figure 9f. Using multiple strips of the fabricated device, the output voltage was improved to 4.3 mV due to the Seebeck effect. Using a fabric thermoelectric generator containing 5 thermoelectric strip units, the maximum output of electrical power (P_{max}) was 12.29 nW at a $\Delta T = 75.2$ K, as shown in Figure 9f [168].

Despite the various advantages of TEGs, their limitations include a low energy conversion efficiency rate, slow technology progression, limited application possibilities, and the requirement of a constant heat source. The fabrication process is also complex because TEGs should be flexible and tightly adhere to the human body.

4.2. Solar Cells on Textiles

In the past decades, next-generation solar cells such as dye-sensitized solar cells (DSSC), organic solar cells (OSC), and perovskite solar cells (PSC) have attracted much attention due to their low cost and high efficiency [171,172,173]. These solar cells are based on light absorption using dye, organic, or perovskite materials that generate mobile electron-hole pairs [174]. The generated electrons and holes are separated and collected into cathodes and anodes, respectively. Recently, fiber-based solar cells have shown great potential for use in textile electronics and represent a new direction in solar energy harvesting since they can be woven into various types of flexible structures at low-cost. However, these devices would benefit from further study into the development of facile fabrication methods for large-scale production and realization of actual products as well as the feasibility of integration in cloth. Several researchers reporting wearable solar cells have primarily focused on inorganic, dye-sensitized, perovskite, and polymer-based solar cells with lightweight, flexible, and ease of manufacture [38,99,175,176].

Recent studies on dye-sensitized photovoltaic fibers have used primarily copper, steel wire, or polymer fibers coated with an indium tin oxide (ITO) electrode [177,178]. Cai et al. demonstrated a new method of fabricating 0.3 mM cis-bis(isothiocyanato)bis(2,2′-bipyridyl-4,4′-dicarboxylato)-ruthenium(II) bis-tetrabutylammonium (N719)-absorbed CNT fibers as a working electrode and CNT fibers coated with poly-vinylidene fluoride (PVDF) as a counter electrode for dye-sensitized photovoltaic application [178]. The fiber-based solar cell was integrated into textiles via conventional weaving techniques. The left side of Figure 10a shows a schematic illustration of the photovoltaic wire, while the middle image in Figure 10a shows an SEM image of the twisted CNT/N719 and CNT/PVDF fibers [178]. The open circuit voltage and short-circuit current density of the fiber-based dye-sensitized solar cell were 0.69 V and 9.84 mA·cm^{-2}, respectively, with a power conversion efficiency of 3.90%, as shown in the right image in Figure 10a.

Zhang et al. demonstrated CdSe NWs and CNT yarn-based photovoltaics, with power conversion efficiency up to 3% [179]. The left image in Figure 10b presents a schematic illustration of fiber-based CdSe-grafted Ti wire fabricated via chemical vapor deposition (CVD) with twisted CNT yarn. The SEM image in Figure 10b shows that the twisted CdSe-Ti wire and CNT fiber make smooth contact without the addition of electrolyte due to the flexibility of the yarn. The performance of CNT yarn and CdSe-Ti wire heterojunction with polysulfide electrodes was characterized by measuring the J-V curve of the fiber cells. The open-circuit voltage (V_{oc}), short-circuit current density (J_{sc}), fill factor (FF), and efficiency were 0.65 V, 8.25 mA·cm^{-2}, 55%, and 2.9%, respectively.

Recent successes in perovskite ($CH_3NH_3PbX_3$, X = Cl, Br, I) solar cells provide encouragement for flexible solar cells in wearable electronics [180,181]. Perovskite materials have many attractive properties including broad spectral absorption and high carrier mobility. The efficiency of thin film photovoltaic cells using perovskite materials have been recorded at up to 20% [182,183]. Some attempts have been made to fabricate fiber-based perovskite solar cells for wearable electronics [38,184]; however, it is difficult to integrate the fibers into clothes due to the insufficient flexibility and low surface area of perovskite materials. To solve this issue, Li et al. described a CNT fiber-supported double-twisted perovskite solar cell featuring high stability where the perovskite nanocrystals on the CNT fiber sustain the interfacial adhesion under stress [38]. The left image in Figure 10d shows a schematic illustration of the CNT fiber-supported double-twisted perovskite solar cell using twisted n-TiO_2 added CNT, meso-TiO_2, $CH_3NH_3PbI_{3-x}Cl_x$, P3HT/SWNT, and an Ag nanowire network [38]. The P3HT/SWNT was used as a hole transport layer, and the Ag nanowire was added to increase the surface area between electrodes. The fabricated perovskite solar cell fiber can be seen in the middle of Figure 10d; the optimized maximum power conversion efficiency (PCE) was 3.03%. Although the performance of the abovementioned textile-based photovoltaic harvesters are, some limitations such as mass production, high cost, and low energy conversion efficiency remained as obstacles to utilization in real fabrics. Further research into textile-based solar cells is needed for the realization of future wearable solar cells.

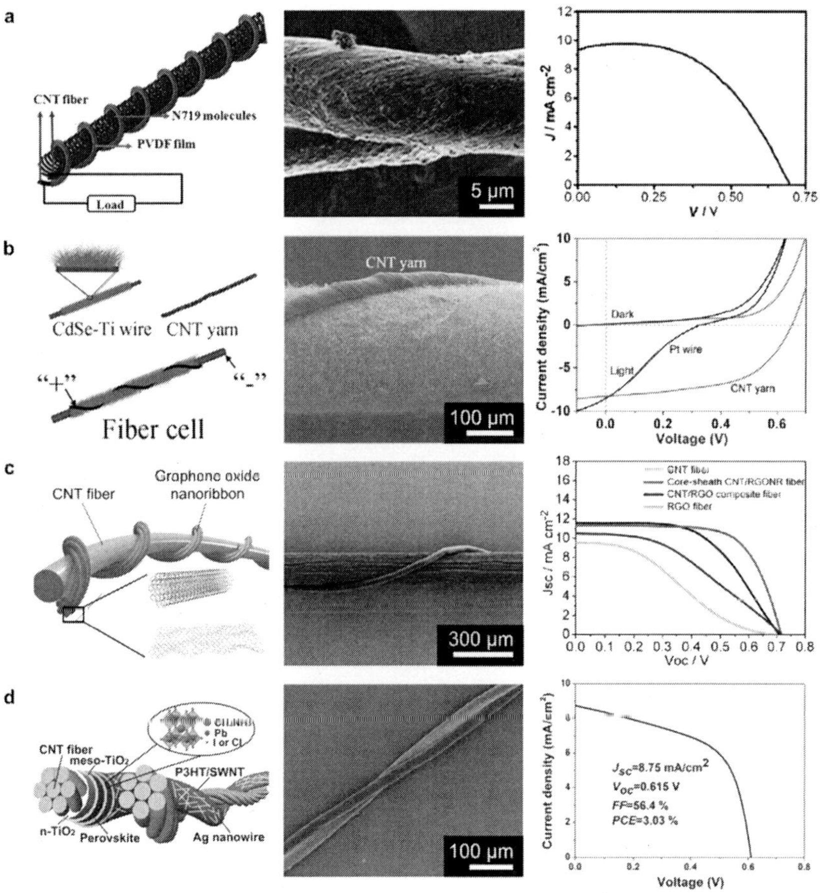

Figure 10. (**a**) Schematic (**left**) and SEM image (**middle**) of the twisted CNT/N719 working electrode and carbon nanotube/poly-vinylidene fluoride (CNT/PVDF) counter electrode. A typical density–voltage (J–V) curve (**right**) with improved performance after modification of the working electrode; (**b**) Schematic fabrication process (**left**) and SEM image (**middle**) of photo electrochemical cells with a primary electrode

of Ti wire, light absorption layer of CdSe nanowire, and counter electrode of CNT. J–V curves under dark and light conditions of different fiber cells using a CNT yarn and Pt wire counter electrode; (**c**) Schematic (**left**) and SEM image (**middle**) of wire shaped dye-sensitized solar cells with twisted CNT fiber. Current density-voltage curves for a wire-shaped DSC (**right**); (**d**) Schematic diagram of double twisted solar cell supported by the CNT fiber and the chemical structure of the perovskite material (**left**). The SEM image shows twisted fibrous solar cells (**middle**). The current J–V curve is shown for electrical performance (**right**). Reprinted from [38,99,175,178] with permissions by Royal Society of Chemistry, Copyright 2012, 2012 and by ACS Publications, Copyright 2013 and by John Wiley & Sons, Copyright 2015.

4.3. Fabric-Based Supercapacitors

Supercapacitors (SCs) have attracted significant attention as outstanding storage components due to their high power density, operational safety, long cycle life, and fast charge-discharge rates [185,186,187]. Depending on the mechanism of charge storage, SCs can be categorized into two types: the Electrical Double Layer Capacitor (EDLC) and pseudocapacitor. EDLCs are typically fabricated using two electrodes separated by a carbon-based electrolyte and an electrical double layer is formed between the electrodes and electrolyte based dielectric. Pseudocapacitors function based on reversible faradaic-type charging and discharging. The supercapacitor is made of conducting polymers and transition metal oxides, and the mechanism of energy storage depends on the redox reactions between materials. To integrate these supercapacitors into textiles, their components, including electrodes and current collectors, will have to be flexible.

Recently, Yang et al. demonstrated a pseudocapacitor-based supercapacitor with improved redox reaction and good reversibility

[188]. To further improve the performance, one study introduced a hybrid architecture using MnO_2-coated hydrogenated zinc oxide nanowires (HZnO) grown on a flexible carbon cloth [189,190]. High electrochemical performance was exhibited by amorphous ZnO-doped MnO_2 core-shell nanocable (HZM) electrodes with a polyvinyl alcohol (PVA)/LiCl neutral electrolyte. Figure 11a shows a schematic illustration of the synthesis process of core-shell structures and an SEM image of HZM. The cyclic voltammetry (CV) measurement shown in Figure 11b indicates that the HZnO electrode had a higher free carrier concentration and better electrical stability when compared to untreated ZnO and air-annealed ZnO electrodes which is attributed to the large surface area of the HZnO electrode.

This improvement was further explained by the insertion of hydrogen acting as a shallow donor for the neutral and electrically-inactive oxygen vacancies in n-type ZnO. Figure 11c shows the combination of the energy harvesting module, dye-sensitized solar cells (DSSCs), and energy storage component that could illuminate a commercial blue LED from charged SCs when the solar source is off. Figure 11d shows a significant initial drop in the leakage current; this measurement stabilized after 4500 s, indicating the presence of few impurities in the electrode and electrolyte. This research contributed to both the fundamental advancement of this technology as well as the advancement of potential of SCs applications.

Zhang et al. also developed a flexible energy fiber composed of titania nanotube-modified Ti wire and MWCNT that can store energy as well as harvest solar energy [191]. P3HT: 1-(3-methoxycarbonyl) propyl-1-phenyl [6,6]C_{61} (PCBM) and PEDOT:PSS with MWCNTs were used as photoactive materials and a PVA/H_3PO_4 gel electrolyte was utilized for capacitor storage, as shown in Figure 11e. When the P3HT: PCBM heterojunction generates excitons and separate carriers (electrons and holes), the carriers are transported through MWCNTs and PEDOT:PSS, and then stored in the energy storage area. Figure 11f shows the energy fibers that were successfully woven into textiles. Figure 11g describes the charge-discharge curves, which indicate a fast charging process. The maximum efficiency and energy storage capacity of the fibers are ~1% and ~65%, respectively. These advances represent various new opportunities for the future of wearable electronics.

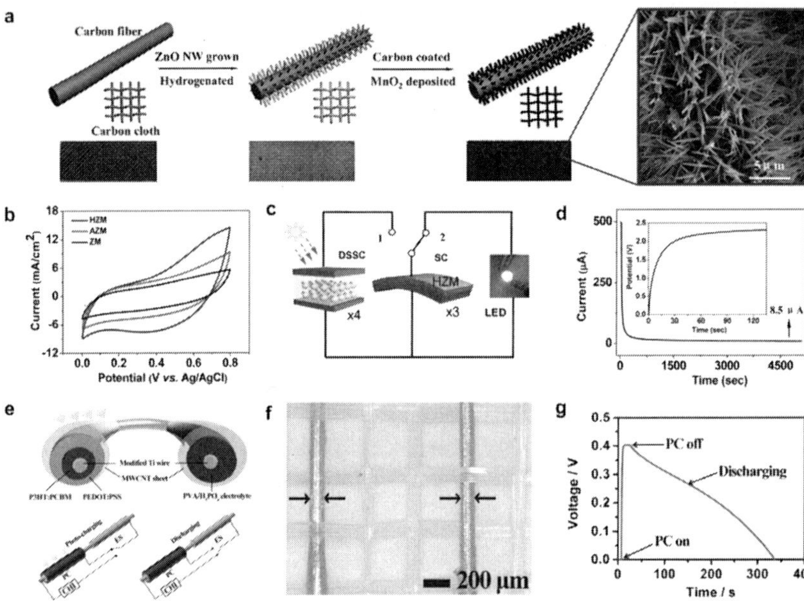

Figure 11. (**a**) Schematic fabrication process and SEM image of a hydrogenated single crystal ZnO (HZM) and amorphous ZnO-doped MnO_2 core-shell electrodes; (**b**) Current-potential characterization with different electrodes with scan rates of 100 mV/s; (**c**) Design of a stand-alone self-powered system consisting of dye-sensitized solar cells (DSSCs), SCs, and LEDs; (**d**) Leakage current characteristics of three SCs in series and voltage profile (inset) charged by four DSSCs in series; (**e**) Schematic illustration and circuit connection mechanism in the process of charging and discharging of photovoltaic conversion (PC) and energy storage (ES) with titania nanotube-modified Ti wire and aligned MWCNT sheet as two electrodes; (**f**) Photography of energy fibers woven with each other to form flexible textiles; (**g**) Charging-discharging curve

with a current of 0.1 μA during the discharging process with the ES part. Reproduced from [188,191] with permissions by ACS Publications, Copyright 2013 and by John Wiley & Sons, Copyright 2014.

5. CURRENT LIMITATIONS OF WEARABLE E-TEXTILES

Aforementioned discussion on the wearable E-textiles presents them as a good futuristic candidate in daily life of human outfits. However, for the true realization of wearable E-textiles, some basic issues must be addressed.

One of the main limitations of wearable E-textiles which require more concern is toxicity due to hazards materials used in electronic components. If, we integrate the electronic components in the wearable textiles they must be safe and non-toxic. Electronic components contain toxic material for example batteries have toxic and liquid electrolyte material like Li-ion batteries and tetraethylammonium tetrafluoroborate in acetonitrile [186,192]. Therefore, garments integrated with these electronic components are not suitable to wear because they will be in directly contact with skin. To resolve this issue some reports are available to use a non-toxic material. Jost et al. have fabricated flexible and durable textile electrodes for supercapacitors by using non-toxic electrolyte material (sodium and lithium sulfate) [193]. The as prepared fabric electrodes have shown high areal capacitance but the problem with these fabric electrodes is the requirement of an integrated device to seal the liquid electrolyte to avoid its leakage. Further, a non-toxic electrolyte based wearable supercapacitor has been suggested by pasta et al. in their well-arranged study [194]. They have reported a simple fabrication technique for supercapacitor showing high specific capacitance and cycling stability. But the drawback of as prepared supercapacitor is the hydrogen evolution overvoltage prevents the operative voltage range as similarly reported by Prosini et al. [195]. Several other reports are also available on non-toxic electrolyte materials. However, further improvements are required in order to use these components in wearable textiles and toxicity issue should be carefully monitored before employ them in wearable fabrics.

Other drawback of these wearable E-textiles is related to their multiple uses, i.e., washability and durability. To the best of our knowledge, there is no study on multiple washable textiles integrated with electronic components. In general, multiple washing and special care is required to worn these garments in daily life. Also, the long term usage of these wearable E-textiles is again a matter of concern. It is required to know that the repairing as well as exchanging of electronic device is possible during their long term use by wearer. The other challenge for researchers is to focus on the packaging of electronic components and device performance for their usage over the longer time. Due to crack and poor bending properties of fiber–based electronics, failure of packaging is observed [196]. These wearable electronic fabrics must hold their sensing, actuating and responding properties in qualitative way during their everyday wear and off-course maintain comfort fit to human body. This can be achieved by replacing hard printed circuit board to flexible electronics.

For the practically implementation of these wearable fabrics the mass production is required and for this an interdisciplinary work must be involved between electronic as well as textile peoples. But to the best of knowledge there is a lack of combination of electronics and textiles industries. Thus, for future developments of E-textiles it is required that the experts of these two firms must work together. Other issues include the biocompatibility of these textiles, development of sophisticated multi-functionalities, stable performance of devices and light weight fabrics for future applications. Thus, any realistic commercial implementation can be achieved after addressing all pointed issues.

6. CONCLUSIONS AND PERSPECTIVES

In this review article, a brief overview of textile electronics is presented based on electronic components, in which the various types of conductive fibers and fabrication techniques are summarized using different materials such as polymers, CNTs, graphene, and metal nanowires. Fiber-based electronic components (transistors, LEDs, and sensors) have also been described. Furthermore, textile-based energy harvesting and storage devices were explained for energy applications in textile electronics. Although considerable performance in textile-based electronic devices has already been achieved, further efforts to improve performance are necessary. A significant body of theoretical and experimental research has been carried out to understand the mechanism and characteristics of E-textiles in wearable electronics. To commercialize textile-based devices in wearable electronics, some important issues will have to be addressed, including mass production, integration into clothes, non toxic technology, and long-term usage. If the E-textiles can overcome the aforementioned issues, a new era in wearable electronics will begin.

ACKNOWLEDGMENTS

This work was supported by the LG Yonam Foundation. We are thankful to the LG group for the support based on Academic Support Programs of LG Yonam Foundation.

AUTHOR CONTRIBUTIONS

All authors discussed the contents of the review manuscript. V.K. wrote the text of the manuscript. S.L., J.H. and J.L. made the Figures for paper. C.M. also contributed to write the manuscript and revised by V.K. and S.L. Modifications and the proof reading of the manuscript has been done by J.S., S.L., J.H., S.L. and V.K. The manuscript is prepared under the supervision of T.L.

REFERENCES

1. Csapó, Á.; Wersényi, G.; Nagy, H.; Stockman, T. A survey of assistive technologies and applications for blind users on mobile platforms: A review and foundation for research. J. Multimodal User Interfaces 2015.

2. Lewis, Z.H.; Lyons, E.J.; Jarvis, J.M.; Baillargeon, J. Using an electronic activity monitor system as an intervention modality: A systematic review. BMC Public Health 2015, 15, 585–600.

3. Vanveerdeghem, P.; Van Torre, P.; Thielens, A.; Knockaert, J.; Joseph, W.; Rogier, H. Compact personal distributed wearable exposimeter. IEEE Sens. J. 2015, 15, 4393–4401.

4. Yun, J.Y.; Hong, W.G.; Choi, N.-J.; Kim, H.B.; Jun, Y.; Lee, H.-K. Ultrasensitive and highly selective graphene-based single yarn for use in wearable gas sensor. Sci. Rep. 2015, 5, 10904–10910.

5. Bai, P.F.; Hayes, R.A.; Jin, M.; Shui, L.; Yi, Z.C.; Wang, L.; Zhang, X.; Zhou, G. Review of paper-like display technologies (invited review). Prog. Electromagn. Res. 2014, 147, 95–116.

6. Patel, S.; Park, H.; Bonato, P.; Chan, L.; Rodgers, M. A review of wearable sensors and systems with application in rehabilitation. J. Neuroeng. Rehabil. 2012, 9, 1–17.

7. Bell, A.G. Phone and Other Inventions. Available online: http://www.carnetdevol.org/Bell/Phone.html (accessed on 3 August 2015).

8. Tropp, H. Computer report VII: The effervescent years: A retrospective: Behind the computer revolution lay the vision and perseverance of a handful of pioneers committed to open information exchange. IEEE Spectr. 1974, 11, 70–79.

9. Goldstine, H.H.; Goldstine, A. The electronic numerical integrator and computer (ENIAC). IEEE Ann. Hist. Comput. 1996, 18, 10–16.

10. Myslewski, R. File:Ibm pcjr with display.jpg. Available online: https://commons.wikimedia.org/wiki/File:Ibm_pcjr_with_display.jpg (accessed on 14 July 2015).

11. Gilbert, D. Mobile Phone Celebrates 40th Birthday. Available online: http://www.ibtimes.co.uk/mobile-phone-40-years-motorola-dynatac-history-453057 (accessed on 3 August 2015).

12. Steven, S. Epson-hx-20. Available online: https://upload.wikimedia.org/wikipedia/commons/8/8d/Epson-hx-20.jpg (accessed on 14 July 2015).

13. Mann, S. Smart clothing: The wearable computer and wearcam. Pers. Technol. 1997, 1, 21–27.

14. Ritchie, R. History of iPhone: Apple reinvents the phone. Available online: http://www.imore.com/history-iphone-2g (accessed on 3 August 2015).

15. Ackerman, E. Google gets in your face. IEEE Spectr. 2013, 50, 26–29.

16. Apple. Apple Watch. Available online: http://www.apple.com/kr/watch/gallery/ (accessed on 3 August 2015).

17. Popat, K.A.; Sharma, P. Wearable computer applications a future perspective. Int. J. Eng. Innov. Technol. 2013, 3, 213–217.

18. Someya Group Organic Transistor Lab Press Kits. Available online: http://www.ntech.t.u-tokyo.ac.jp/en/press/press_for_media/index.html (accessed on 3 August 2015).

19. Caron, N. Meet the Rest of the Cast for Fox's Minority Report TV series. Available online: http://www.blastr.com/2015-3-3/meet-rest-cast-foxs-minority-report-tv-series (accessed on 3 August 2015).

20. Lee, J.; Kwon, H.; Seo, J.; Shin, S.; Koo, J.H.; Pang, C.; Son, S.; Kim, J.H.; Jang, Y.H.; Kim, D.E.; Lee, T. Conductive fiber-based ultrasensitive textile pressure sensor for wearable electronics. Adv. Mater. 2015, 27, 2433–2439.

21. Seesaard, T.; Lorwongtragool, P.; Kerdcharoen, T. Development of fabric-based chemical gas sensors for use as wearable electronic noses. Sensors 2015, 15, 1885–1902.

22. Skrzetuska, E.; Puchalski, M.; Krucińska, I. Chemically driven printed textile sensors based on graphene and carbon nanotubes. Sensors 2014, 14, 16816–16828.

23. Lanata, A.; Valenza, G.; Nardelli, M.; Gentili, C.; Scilingo, E.P. Complexity index from a personalized wearable monitoring system for assessing remission in mental health. IEEE J. Biomed. Health Inform. 2015, 19, 132–139.

24. Zheng, Y.-L.; Ding, X.-R.; Poon, C.C.Y.; Lo, B.P.L.; Zhang, H.; Zhou, X.-L.; Yang, G.-Z.; Zhao, N.; Zhang, Y.-T. Unobtrusive sensing and wearable devices for health informatics. IEEE Trans. Biomed. Eng. 2014, 61, 1538–1554.

25. Zheng, N.; Wu, Z.; Lin, M.; Yang, L.T. Enhancing battery efficiency for pervasive health-monitoring systems based on electronic textiles. IEEE Trans. Inf. Technol. Biomed. 2010, 14, 350–359.

26. Paradiso, R.; Loriga, G.; Taccini, N. A wearable health care system based on knitted integrated sensors. IEEE Trans. Inf. Technol. Biomed. 2005, 9, 337–344.

27. Quinlan, M.; Mayhew, C. The effects of outsourcing on occupational health and safety: A comparative study of factory-based workers and outworkers in the Australian clothing industry. Int. J. Health Serv. 1999, 29, 83–107.

28. Rofouei, M.; Potkonjak, M.; Sarrafzadeh, M. Energy efficient collaborative sensing-based design: Soft keyboard case study. IEEE Trans. Hum. Mach. Syst. 2014, 44, 115–124.

29. Yoo, H.-J. Your heart on your sleeve: Advances in textile-based electronics are weaving computers right into the clothes we wear. IEEE Solid State Circuits Mag. 2013, 5, 59–70.

30. Li, W.; Zhu, L.; Gu, J.; Liu, H. Microwave absorption properties of fabric coated absorbing material using modified carbonyl iron power. Compos. Part B Eng. 2011, 42, 626–630.

31. Moro, R.; Agneessens, S.; Rogier, H.; Dierck, A.; Bozzi, M. Textile microwave components in substrate integrated waveguide technology. IEEE Trans. Microw. Theory Tech. 2015, 63, 422–432.

32. Castano, L.M.; Flatau, A.B. Smart fabric sensors and E-textile technologies: A review. Smart Mater. Struct. 2014, 23, 053001–053027.

33. Stoppa, M.; Chiolerio, A. Wearable electronics and smart textiles: A critical review. Sensors 2014, 14, 11957–11992.

34. Zięba, J.; Frydrysiak, M. Textronics–electrical and electronic textiles. Sensors for breathing frequency measurement. Fibres Text. East. Eur. 2006, 14, 43–48.

35. Sun, H.; You, X.; Yang, Z.; Deng, J.; Peng, H. Winding ultrathin, transparent, and electrically conductive carbon nanotube sheets into high-performance fiber-shaped dye-sensitized solar cells. J. Mater. Chem. A 2013, 1, 12422–12425.

36. Weintraub, B.; Wei, Y.; Wang, Z.L. Optical fiber/nanowire hybrid structures for efficient three-dimensional dye-sensitized solar cells. Angew. Chem. Int. Ed. 2009, 48, 8981–8985.

37. Liu, D.; Zhao, M.; Li, Y.; Bian, Z.; Zhang, L.; Shang, Y.; Xia, X.; Zhang, S.; Yun, D.; Liu, Z.; Cao, A.; Huang, C. Solid-state, polymer-based fiber solar cells with carbon nanotube electrodes. ACS Nano 2012, 6, 11027–11034.

38. Li, R.; Xiang, X.; Tong, X.; Zou, J.; Li, Q. Wearable double-twisted fibrous perovskite solar cell. Adv. Mater. 2015, 27, 3831–3835.

39. Lee, J.B.; Subramanian, V. Weave patterned organic transistors on fiber for E-textiles. IEEE Trans. Electron Devices 2005, 52, 269–275.

40. Tao, X.; Koncar, V.; Dufour, C. Geometry pattern for the wire organic electrochemical textile transistor. J. Electrochem. Soc. 2011, 158, H572–H577.

41. Kylberg, W.; de Castro, F.A.; Chabrecek, P.; Sonderegger, U.; Chu, B.T.-T.; Nüesch, F.; Hany, R. Woven electrodes for flexible organic photovoltaic cells. Adv. Mater. 2011, 23, 1015–1019.

42. Zhang, Z.; Guo, K.; Li, Y.; Li, X.; Guan, G.; Li, H.; Luo, Y.; Zhao, F.; Zhang, Q.; Wei, B.; Pei, Q.; Peng, H. A colour-tunable, weavable fibre-shaped polymer light-emitting electrochemical cell. Nat. Photonics 2015, 9, 233–238.

43. Vohra, V.; Giovanella, U.; Tubino, R.; Murata, H.; Botta, C. Electroluminescence from conjugated polymer electrospun nanofibers in solution processable organic light-emitting diodes. ACS Nano 2011, 5, 5572–5578.

44. Bedeloglu, A.; Demir, A.; Bozkurt, Y.; Sariciftci, N.S. A flexible textile structure based on polymeric photovoltaics using transparent cathode. Synth. Met. 2009, 159, 2043–2048.

45. Jost, K.; Dion, G.; Gogotsi, Y. Textile energy storage in perspective. J. Mater. Chem. A 2014, 2, 10776–10787.

46. Wang, Z.L. Towards self-powered nanosystems: From nanogenerators to nanopiezotronics. Adv. Funct. Mater. 2008, 18, 3553–3567.

47. Das, T.K.; Prusty, S. Review on conducting polymers and their applications. Polym. Plast. Technol. Eng. 2012, 51, 1487–1500.

48. Wegner, G. Polymers with metal-like conductivity—A review of their synthesis, structure and properties. Angew. Chem. Int. Ed. Engl. 1981, 20, 361–381.

49. Patil, A.O.; Heeger, A.J.; Wudl, F. Optical properties of conducting polymers. Chem. Rev. 1988, 88, 183–200.

50. Gerard, M.; Chaubey, A.; Malhotra, B.D. Application of conducting polymers to biosensors. Biosens. Bioelectron. 2002, 17, 345–359.

51. Janata, J.; Josowicz, M. Conducting polymers in electronic chemical sensors. Nat. Mater. 2003, 2, 19–24.

52. Svirskis, D.; Travas-Sejdic, J.; Rodgers, A.; Garg, S. Electrochemically controlled drug delivery based on intrinsically conducting polymers. J. Controlled Release 2010, 146, 6–15.

53. Pernaut, J.-M.; Reynolds, J.R. Use of conducting electroactive polymers for drug delivery and sensing of bioactive molecules. A redox chemistry approach. J. Phys. Chem. B 2000, 104, 4080–4090.

54. Tallman, D.E.; Spinks, G.; Dominis, A.; Wallace, G.G. Electroactive conducting polymers for corrosion control. J. Solid State Electrochem. 2001, 6, 73–84.

55. Hohnholz, D.; Okuzaki, H.; MacDiarmid, A.G. Plastic electronic devices through line patterning of conducting polymers. Adv. Funct. Mater. 2005, 15, 51–56.

56. Green, R.A.; Lovell, N.H.; Wallace, G.G.; Poole-Warren, L.A. Conducting polymers for neural interfaces: Challenges in developing an effective long-term implant. Biomaterials 2008, 29, 3393–3399.

57. Olmedo, L.; Hourquebie, P.; Jousse, F. Microwave absorbing materials based on conducting polymers. Adv. Mater. 1993, 5, 373–377.

58. Malinauskas, A. Electrocatalysis at conducting polymers. Synth. Met. 1999, 107, 75–83.

59. Kingsborough, R.P.; Swager, T.M. Electrocatalytic conducting polymers: Oxygen reduction by a polythiophene-cobalt salen hybrid. Chem. Mater. 2000, 12, 872–874.

60. Shirakawa, H.; Louis, E.J.; MacDiarmid, A.G.; Chiang, C.K.; Heeger, A.J. Synthesis of electrically conducting organic polymers: Halogen derivatives of polyacetylene, (CH)x. J. Chem. Soc. Chem. Commun. 1977, 578–580.

61. Schreuder-Gibson, H.L.; Truong, Q.; Walker, J.E.; Owens, J.R.; Wander, J.D.; Jones, W.E. Chemical and biological protection and detection in fabrics for protective clothing. MRS Bull. 2003, 28, 574–578.

62. Cochrane, C.; Koncar, V.; Lewandowski, M.; Dufour, C. Design and development of a flexible strain sensor for textile structures based on a conductive polymer composite. Sensors 2007, 7, 473–492.

63. De Rossi, D.; Carpi, F.; Lorussi, F.; Mazzoldi, A.; Paradiso, R.; Scilingo, E.P.; Tognetti, A. Electroactive fabrics and wearable biomonitoring devices. AUTEX Res. J. 2003, 3, 180–185.

64. Smith, D. Smart clothes and wearable technology. AI Soc. 2007, 22, 1–3.

65. Park, S.; Jayaraman, S. Smart textiles: Wearable electronic systems. MRS Bull. 2003, 28, 585–591.

66. Kim, M.S.; Kim, H.K.; Byun, S.W.; Jeong, S.H.; Hong, Y.K.; Joo, J.S.; Song, K.T.; Kim, J.K.; Lee, C.J.; Lee, J.Y. PET fabric/polypyrrole composite with high electrical conductivity for EMI shielding. Synth. Met. 2002, 126, 233–239.

67. Varesano, A.; Dall'Acqua, L.; Tonin, C. A study on the electrical conductivity decay of polypyrrole coated wool textiles. Polym. Degrad. Stab. 2005, 89, 125–132.

68. Ding, Y.; Invernale, M.A.; Sotzing, G.A. Conductivity trends of PEDOT-PSS impregnated fabric and the effect of conductivity on electrochromic textile. ACS Appl. Mater. Interfaces 2010, 2, 1588–1593.

69. Gospodinova, N.; Terlemezyan, L. Conducting polymers prepared by oxidative polymerization: Polyaniline. Prog. Polym. Sci. 1998, 23, 1443–1484.

70. Jianming, J.; Wei, P.; Shenglin, Y.; Guang, L. Electrically conductive PANI-DBSA/Co-PAN composite fibers prepared by wet spinning. Synth. Met. 2005, 149, 181–186.

71. Dallas, P.; Niarchos, D.; Vrbanic, D.; Boukos, N.; Pejovnik, S.; Trapalis, C.; Petridis, D. Interfacial polymerization of pyrrole and in situ synthesis of polypyrrole/silver nanocomposites. Polymer 2007, 48, 2007–2013.

72. Müller, D.; Rambo, C.R.; Recouvreux, D.O.S.; Porto, L.M.; Barra, G.M.O. Chemical in situ polymerization of polypyrrole on bacterial cellulose nanofibers. Synth. Met. 2011, 161, 106–111.

73. Xu, R.; Wei, J.; Guo, F.; Cui, X.; Zhang, T.; Zhu, H.; Wang, K.; Wu, D. Highly conductive, twistable and bendable polypyrrole–carbon nanotube fiber for efficient supercapacitor electrodes. RSC Adv. 2015, 5, 22015–22021.

74. Zhang, Y.; Rutledge, G.C. Electrical conductivity of electrospun polyaniline and polyaniline-blend fibers and mats. Macromolecules 2012, 45, 4238–4246.

75. Balint, R.; Cassidy, N.J.; Cartmell, S.H. Conductive polymers: Towards a smart biomaterial for tissue engineering. Acta Biomater. 2014, 10, 2341–2353.

76. Moulé, A.J.; Jung, M.-C.; Rochester, C.W.; Tress, W.; LaGrange, D.; Jacobs, I.E.; Li, J.; Mauger, S.A.; Rail, M.D.; Lin, O.; Bilsky, D.J.; Qi, Y.; Stroeve, P.; Berben, L.A.; Riede, M. Mixed interlayers at the interface between PEDOT:PSS and conjugated polymers provide charge transport control. J. Mater. Chem. C 2015, 3, 2664–2676.

77. Pomfret, S.J.; Adams, P.N.; Comfort, N.P.; Monkman, A.P. Electrical and mechanical properties of polyaniline fibres produced by a one-step wet spinning process. Polymer 2000, 41, 2265–2269.

78. Huang, K.; Wan, M.; Long, Y.; Chen, Z.; Wei, Y. Multi-functional polypyrrole nanofibers via a functional dopant-introduced process. Synth. Met. 2005, 155, 495–500.

79. Wallace, G.G.; Campbell, T.E.; Innis, P.C. Putting function into fashion: Organic conducting polymer fibres and textiles. Fibers Polym. 2007, 8, 135–142.

80. Chacko, A.P.; Hardaker, S.S.; Gregory, R.V.; Samuels, R.J. Viscoelastic characterization of concentrated polyaniline solutions: New insights into conductive polymer processing. Synth. Met. 1997, 84, 41–44.

81. Mottaghitalab, V.; Spinks, G.M.; Wallace, G.G. The development and characterisation of polyaniline—Single walled carbon nanotube composite fibres using 2-acrylamido-2 methyl-1-propane sulfonic acid (AMPSA) through one step wet spinning process. Polymer 2006, 47, 4996–5002.

82. Takahashi, T.; Ishihara, M.; Okuzaki, H. Poly(3,4-ethylenedioxythiophene)/poly(4-styrenesulfonate) microfibers. Synth. Met. 2005, 152, 73–76.

83. Yun, Y.J.; Hong, W.G.; Kim, W.-J.; Jun, Y.; Kim, B.H. A novel method for applying reduced graphene oxide directly to electronic textiles from yarns to fabrics. Adv. Mater. 2013, 25, 5701–5705.

84. Meng, Y.; Zhao, Y.; Hu, C.; Cheng, H.; Hu, Y.; Zhang, Z.; Shi, G.; Qu, L. All-graphene core-sheath microfibers for all-solid-state, stretchable fibriform supercapacitors and wearable electronic textiles. Adv. Mater. 2013, 25, 2326–2331.

85. Jang, E.Y.; Kang, T.J.; Im, H.; Baek, S.J.; Kim, S.; Jeong, D.H.; Park, Y.W.; Kim, Y.H. Macroscopic single-walled-carbon-nanotube fiber self-assembled by dip-coating method. Adv. Mater. 2009, 21, 4357–4361.

86. Chen, T.; Cai, Z.; Yang, Z.; Li, L.; Sun, X.; Huang, T.; Yu, A.; Kia, H.G.; Peng, H. Nitrogen-doped carbon nanotube composite fiber with a core-sheath structure for novel electrodes. Adv. Mater. 2011, 23, 4620–4625.

87. Hu, L.; Pasta, M.; Mantia, F.L.; Cui, L.; Jeong, S.; Deshazer, H.D.; Choi, J.W.; Han, S.M.; Cui, Y. Stretchable, porous, and conductive energy textiles. Nano Lett. 2010, 10, 708–714.

88. Lee, H.M.; Choi, S.-Y.; Jung, A.; Ko, S.H. Highly conductive aluminum textile and paper for flexible and wearable electronics. Angew. Chem. Int. Ed. 2013, 52, 7718–7723.

89. Park, M.; Im, J.; Shin, M.; Min, Y.; Park, J.; Cho, H.; Park, S.; Shim, M.-B.; Jeon, S.; Chung, D.-Y.; et al. Highly stretchable electric circuits from a composite material of silver nanoparticles and elastomeric fibres. Nat. Nanotechnol. 2012, 7, 803–809.

90. Lee, S.; Shin, S.; Lee, S.; Seo, J.; Lee, J.; Son, S.; Cho, H.J.; Algadi, H.; Al-Sayari, S.; Kim, D.E.; Lee, T. Ag nanowire reinforced highly stretchable conductive fibers for wearable electronics. Adv. Funct. Mater. 2015, 25, 3114–3121.

91. Wu, A.S.; Nie, X.; Hudspeth, M.C.; Chen, W.W.; Chou, T.-W.; Lashmore, D.S.; Schauer, M.W.; Tolle, E.; Rioux, J. Strain rate-dependent tensile properties and dynamic electromechanical response of carbon nanotube fibers. Carbon 2012, 50, 3876–3881.

92. Koziol, K.; Vilatela, J.; Moisala, A.; Motta, M.; Cunniff, P.; Sennett, M.; Windle, A. High-performance carbon nanotube fiber. Science 2007, 318, 1892–1895.

93. Chen, H.; Müller, M.B.; Gilmore, K.J.; Wallace, G.G.; Li, D. Mechanically strong, electrically conductive, and biocompatible graphene paper. Adv. Mater. 2008, 20, 3557–3561.

94. Baughman, R.H.; Zakhidov, A.A.; de Heer, W.A. Carbon nanotubes—The route toward applications. Science 2002, 297, 787–792.

95. Kaushik, V.; Sharma, H.; Girdhar, P.; Shukla, A.K.; Vankar, V.D. Structural modification and enhanced electron emission from multiwalled carbon nanotubes grown on Ag/Fe catalysts coated Si-substrates. Mater. Chem. Phys. 2011, 130, 986–992.

96. Cao, A.; Dickrell, P.L.; Sawyer, W.G.; Ghasemi-Nejhad, M.N.; Ajayan, P.M. Super-compressible foamlike carbon nanotube films. Science 2005, 310, 1307–1310.

97. Jakubinek, M.B.; Johnson, M.B.; White, M.A.; Jayasinghe, C.; Li, G.; Cho, W.; Schulz, M.J.; Shanov, V. Thermal and electrical conductivity of array-spun multi-walled carbon nanotube yarns. Carbon 2012, 50, 244–248.

98. Ericson, L.M.; Fan, H.; Peng, H.; Davis, V.A.; Zhou, W.; Sulpizio, J.; Wang, Y.; Booker, R.; Vavro, J.; Guthy, C.; et al. Macroscopic, neat, single-walled carbon nanotube fibers. Science 2004, 305, 1447–1450.

99. Zhang, L.; Shi, E.; Ji, C.; Li, Z.; Li, P.; Shang, Y.; Li, Y.; Wei, J.; Wang, K.; Zhu, H.; et al. Fiber and fabric solar cells by directly weaving carbon nanotube yarns with CdSe nanowire-based electrodes. Nanoscale 2012, 4, 4954–4959.

100. Davis, V.A.; Parra-Vasquez, A.N.G.; Green, M.J.; Rai, P.K.; Behabtu, N.; Prieto, V.; Booker, R.D.; Schmidt, J.; Kesselman, E.; Zhou, W.; et al. True solutions of single-walled carbon nanotubes for assembly into macroscopic materials. Nat. Nanotechnol. 2009, 4, 830–834.

101. Vigolo, B.; Pénicaud, A.; Coulon, C.; Sauder, C.; Pailler, R.; Journet, C.; Bernier, P.; Poulin, P. Macroscopic fibers and ribbons of oriented carbon nanotubes. Science 2000, 290, 1331–1334.

102. Shim, B.S.; Chen, W.; Doty, C.; Xu, C.; Kotov, N.A. Smart electronic yarns and wearable fabrics for human biomonitoring

made by carbon nanotube coating with polyelectrolytes. Nano Lett. 2008, 8, 4151–4157.

103. Zhang, S.; Koziol, K.K.K.; Kinloch, I.A.; Windle, A.H. Macroscopic fibers of well-aligned carbon nanotubes by wet spinning. Small 2008, 4, 1217–1222.

104. Zhao, Y.; Wei, J.; Vajtai, R.; Ajayan, P.M.; Barrera, E.V. Iodine doped carbon nanotube cables exceeding specific electrical conductivity of metals. Sci. Rep. 2011, 1, 83–87.

105. Jiang, K.; Li, Q.; Fan, S. Nanotechnology: Spinning continuous carbon nanotube yarns. Nature 2002, 419, 801–801.

106. Xu, Z.; Zhang, Y.; Li, P.; Gao, C. Strong, conductive, lightweight, neat graphene aerogel fibers with aligned pores. ACS Nano 2012, 6, 7103–7113.

107. Scholz, J.; Nocke, G.; Hollstein, F.; Weissbach, A. Investigations on fabrics coated with precious metals using the magnetron sputter technique with regard to their anti-microbial properties. Surf. Coat. Technol. 2005, 192, 252–256.

108. Hegemann, D.; Amberg, M.; Ritter, A.; Heuberger, M. Recent developments in Ag metallised textiles using plasma sputtering. Mater. Technol. 2009, 24, 41–45.

109. Little, B.K.; Li, Y.; Cammarata, V.; Broughton, R.; Mills, G. Metallization of kevlar fibers with gold. ACS Appl. Mater. Interfaces 2011, 3, 1965–1973.

110. Zhang, Z.; Dai, S.; Blom, D.A.; Shen, J. Synthesis of ordered metallic nanowires inside ordered mesoporous materials through electroless deposition. Chem. Mater. 2002, 14, 965–968.

111. Zrnic, D.; Swatik, D.S. On the resistivity and surface tension of the eutectic alloy of gallium and indium. J. Common Met. 1969, 18, 67–68.

112. Zhu, S.; So, J.-H.; Mays, R.; Desai, S.; Barnes, W.R.; Pourdeyhimi, B.; Dickey, M.D. Ultrastretchable fibers with metallic conductivity using a liquid metal alloy core. Adv. Funct. Mater. 2013, 23, 2308–2314.

113. Atwa, Y.; Maheshwari, N.; Goldthorpe, I.A. Silver nanowire coated threads for electrically conductive textiles. J. Mater. Chem. C 2015, 3, 3908–3912.

114. Horowitz, G. Organic field-effect transistors. Adv. Mater. 1998, 10, 365–377.

115. Muccini, M. A bright future for organic field-effect transistors. Nat. Mater. 2006, 5, 605–613.

116. Małachowski, M.J.; Żmija, J. Organic field-effect transistors. Opto-Electron. Rev. 2010, 18, 121–136.

117. Sun, Y.; Liu, Y.; Zhu, D. Advances in organic field-effect transistors. J. Mater. Chem. 2005, 15, 53–65.

118. Lee, J.B.; Subramanian, V. Organic Transistors on Fiber: A First Step Towards Electronic Textiles. In Proceedings of IEEE International Electron Devices Meeting (IEDM '03), Technical Digest, Washington, DC, USA, 8–10 December 2003; pp. 8.3.1–8.3.4.

119. Garnier, F.; Horowitz, G.; Peng, X.; Fichou, D. An all-organic "soft" thin film transistor with very high carrier mobility. Adv. Mater. 1990, 2, 592–594.

120. Tsumura, A.; Koezuka, H.; Ando, T. Macromolecular electronic device: Field-effect transistor with a polythiophene thin film. Appl. Phys. Lett. 1986, 49, 1210–1212.

121. De Rossi, D. Electronic textiles: A logical step. Nat. Mater. 2007, 6, 328–329.

122. Maccioni, M.; Orgiu, E.; Cosseddu, P.; Locci, S.; Bonfiglio, A. Towards the textile transistor: Assembly and characterization of an organic field effect transistor with a cylindrical geometry. Appl. Phys. Lett. 2006, 89, 143515–143517.

123. Hamedi, M.; Forchheimer, R.; Inganäs, O. Towards woven logic from organic electronic fibres. Nat. Mater. 2007, 6, 357–362.

124. Hamedi, M.; Herlogsson, L.; Crispin, X.; Marcilla, R.; Berggren, M.; Inganäs, O. Fiber-embedded electrolyte-gated field-effect transistors for e-textiles. Adv. Mater. 2009, 21, 573–577.

125. Bonderover, E.; Wagner, S. A Woven Inverter Circuit for E-Textile Applications. IEEE Electron Device Lett. 2004, 25, 295–297.

126. Holonyak, N. News - First LED by the GE Engineer, Nick Holonyak. Available online: http://www.gelighting.com/LightingWeb/emea/news-and-media/news/First-LED-by-the-GE-engineer-Nick-Holonyak.jsp (accessed on 16 July 2015).

127. Tan, J.J.; Zou, C.Q.; Du, S.H.; Tan, J.T. Simulation of MIMO channel characteristics for indoor visible light communication with LEDs. Opt. Int. J. Light Electron Opt. 2014, 125, 44–49.

128. Yu, H.; Deng, D.; Chen, L.; Chen, D.; Zhong, J.; Zhao, H.; Xu, S. Luminescent properties of novel Ca2.89Mg0.11(PO4)2:Eu2+ single-phase white light-emitting phosphor for white LEDs. Ceram. Int. 2015, 41, 3800–3805.

129. Zhang, Z.; Wang, J.; Yin, Y.; Zhou, J.; Zhou, S.; Pan, Y. A novel green-to-yellow emitting phosphor: BaSi2SN2.67:Eu2+ for potential application in UV-LEDs. Opt. Mater. 2013, 35, 1273–1275.

130. Park, S.; Choi, I.-M.; Kim, S.-S.; Kim, S.-M. A portable mid-range localization system using infrared LEDs for visually impaired people. Infrared Phys. Technol. 2014, 67, 583–589.

131. Lourenço, M.A.; Homewood, K.P. Crystalline-silicon-based infrared LEDs and routes to laser diodes. Thin Solid Films 2011, 519, 8441–8445.

132. Tang, C.W.; VanSlyke, S.A. Organic electroluminescent diodes. Appl. Phys. Lett. 1987, 51, 913–915.

133. Misra, A.; Kumar, P.; Kamalasanan, M.N.; Chandra, S. White organic LEDs and their recent advancements. Semicond. Sci. Technol. 2006, 21, R35–R47.

134. O'Connor, B.; An, K.H.; Zhao, Y.; Pipe, K.P.; Shtein, M. Fiber shaped light emitting device. Adv. Mater. 2007, 19, 3897–3900.

135. Yang, H.; Lightner, C.R.; Dong, L. Light-emitting coaxial nanofibers. ACS Nano 2012, 6, 622–628.

136. Wise, K.D. Integrated sensors: Interfacing electronics to a non-electronic world. Sens. Actuators 1981, 2, 229–237.

137. Nese, M.; Seeberg, B.E. Silicon MEMS pressure sensors for aerospace applications. ASME Proc. 2006, 195–198.

138. Lin, L.; Yun, W. MEMS Pressure Sensors for Aerospace Applications. In Proceedings of 1998 IEEE Aerospace Conference, Aspen, CO, USA, 28 March 1998; Volume 1, pp. 429–436.

139. Mishra, V.; Singh, N.; Tiwari, U.; Kapur, P. Fiber grating sensors in medicine: Current and emerging applications. Sens. Actuators Phys. 2011, 167, 279–290.

140. Girão, P.S.; Ramos, P.M.P.; Postolache, O.; Miguel Dias Pereira, J. Tactile sensors for robotic applications. Measurement 2013, 46, 1257–1271.

141. Wu, W.; Liang, L.; Xu, R.; Liu, X.; Qin, Z. The application of fiber grating sensor network in port-machinery monitoring. Opt. Lasers Eng. 2010, 48, 427–434.

142. Kinnunen, M.; Mian, S.Q.; Oinas-Kukkonen, H.; Riekki, J.; Jutila, M.; Ervasti, M.; Ahokangas, P.; Alasaarela, E. Wearable and mobile sensors connected to social media in human well-being applications. Telemat. Inform. 2016, 33, 92–101.

143. Mansoor, M.; Haneef, I.; Akhtar, S.; De Luca, A.; Udrea, F. Silicon diode temperature sensors—A review of applications. Sens. Actuators Phys. 2015, 232, 63–74.

144. Della Lucia, F.; Zambrozi, P., Jr.; Frazatto, F.; Piazzetta, M.; Gobbi, A. Design, fabrication and characterization of SAW pressure sensors for offshore oil and gas exploration. Sens. Actuators Phys. 2015, 222, 322–328.

145. Yamazoe, N.; Shimizu, Y. Humidity sensors: Principles and applications. Sens. Actuators 1986, 10, 379–398.

146. Zhu, Z.-T.; Mason, J.T.; Dieckmann, R.; Malliaras, G.G. Humidity sensors based on pentacene thin-film transistors. Appl. Phys. Lett. 2002, 81, 4643–4645.

147. Merkys, A.J.; Laurinavichius, R.S.; Rupainene, O.J.; Savichene, E.K.; Jaroshius, A.V.; Shvegzhdene, D.V.; Bendoraityte, D.P. The state of gravity sensors and peculiarities of plant growth during different gravitational loads. Adv. Space Res. 1983, 3, 211–219.

148. Pasterkamp, H.; Kraman, S.S.; DeFrain, P.D.; Wodicka, G.R. Measurement of respiratory acoustical signals. Comparison of sensors. Chest 1993, 104, 1518–1525.

149. Turner, R.C.; Fuierer, P.A.; Newnham, R.E.; Shrout, T.R. Materials for high temperature acoustic and vibration sensors: A review. Appl. Acoust. 1994, 41, 299–324.

150. Papadimitriou, C.; Fritzen, C.-P.; Kraemer, P.; Ntotsios, E. Fatigue predictions in entire body of metallic structures from a limited number of vibration sensors using Kalman filtering. Struct. Control Health Monit. 2011, 18, 554–573.

151. Inaba, M.; Hoshino, Y.; Nagasaka, K.; Ninomiya, T.; Kagami, S.; Inoue, H. A Full-Body Tactile Sensor Suit Using Electrically Conductive Fabric and Strings. In Proceedings of the 1996 IEEE/RSJ International Conference on Intelligent Robots and Systems '96 (IROS 96), Osaka, Japan, 4–8 November 1996; Volume 2, pp. 450–457.

152. Shimojo, M.; Namiki, A.; Ishikawa, M.; Makino, R.; Mabuchi, K. A tactile sensor sheet using pressure conductive rubber with electrical-wires stitched method. IEEE Sens. J. 2004, 4, 589–596.

153. Takamatsu, S.; Kobayashi, T.; Shibayama, N.; Miyake, K.; Itoh, T. Fabric pressure sensor array fabricated with die-coating and weaving techniques. Sens. Actuators Phys. 2012, 184, 57–63.

154. Ma, R.; Lee, J.; Choi, D.; Moon, H.; Baik, S. Knitted fabrics made from highly conductive stretchable fibers. Nano Lett. 2014, 14, 1944–1951.

155. Starner, T. Human-powered wearable computing. IBM Syst. J. 1996, 35, 618–629.

156. Wang, X.; Song, J.; Liu, J.; Wang, Z.L. Direct-current nanogenerator driven by ultrasonic waves. Science 2007, 316, 102–105.

157. Paradiso, J.; Starner, T. Energy scavenging for mobile and wireless electronics. IEEE Pervasive Comput. 2005, 4, 18–27.

158. Zhang, M.; Gao, T.; Wang, J.; Liao, J.; Qiu, Y.; Yang, Q.; Xue, H.; Shi, Z.; Zhao, Y.; Xiong, Z.; Chen, L. A hybrid fibers based wearable fabric piezoelectric nanogenerator for energy harvesting application. Nano Energy 2015, 13, 298–305.

159. Bouendeu, E.; Greiner, A.; Smith, P.J.; Korvink, J.G. A low-cost electromagnetic generator for vibration energy harvesting. IEEE Sens. J. 2011, 11, 107–113.

160. Mitcheson, P.D.; Miao, P.; Stark, B.H.; Yeatman, E.M.; Holmes, A.S.; Green, T.C. MEMS electrostatic micropower generator for low frequency operation. Sens. Actuators Phys. 2004, 115, 523–529.

161. Qin, Y.; Wang, X.; Wang, Z.L. Microfibre–nanowire hybrid structure for energy scavenging. Nature 2008, 451, 809–813.

162. Jeong, C.K.; Lee, J.; Han, S.; Ryu, J.; Hwang, G.-T.; Park, D.Y.; Park, J.H.; Lee, S.S.; Byun, M.; Ko, S.H.; et al. A hyper-stretchable elastic-composite energy harvester. Adv. Mater. 2015, 27, 2866–2875.

163. Zhong, J.; Zhang, Y.; Zhong, Q.; Hu, Q.; Hu, B.; Wang, Z.L.; Zhou, J. Fiber-based generator for wearable electronics and mobile medication. ACS Nano 2014, 8, 6273–6280.

164. Jung, S.; Lee, J.; Hyeon, T.; Lee, M.; Kim, D.-H. Fabric-based integrated energy devices for wearable activity monitors. Adv. Mater. 2014, 26, 6329–6334.

165. Lee, S.; Ko, W.; Oh, Y.; Lee, J.; Baek, G.; Lee, Y.; Sohn, J.; Cha, S.; Kim, J.; Park, J.; et al. Triboelectric energy harvester based on wearable textile platforms employing various surface morphologies. Nano Energy 2015, 12, 410–418.

166. Fan, F.-R.; Tian, Z.-Q.; Lin Wang, Z. Flexible triboelectric generator. Nano Energy 2012, 1, 328–334.

167. Zhu, G.; Pan, C.; Guo, W.; Chen, C.-Y.; Zhou, Y.; Yu, R.; Wang, Z.L. Triboelectric-generator-driven pulse electrodeposition for micropatterning. Nano Lett. 2012, 12, 4960–4965.

168. Du, Y.; Cai, K.; Chen, S.; Wang, H.; Shen, S.Z.; Donelson, R.; Lin, T. Thermoelectric fabrics: Toward power generating clothing. Sci. Rep. 2015, 5, 6411–6416.

169. Kim, M.-K.; Kim, M.-S.; Lee, S.; Kim, C.; Kim, Y.-J. Wearable thermoelectric generator for harvesting human body heat energy. Smart Mater. Struct. 2014, 23, 105002–105008.

170. Liang, D.; Yang, H.; Finefrock, S.W.; Wu, Y. Flexible nanocrystal-coated glass fibers for high-performance thermoelectric energy harvesting. Nano Lett. 2012, 12, 2140–2145.

171. O'Regan, B.; Grätzel, M. A low-cost, high-efficiency solar cell based on dye-sensitized colloidal TiO2 films. Nature 1991, 353, 737–740.

172. Kippelen, B.; Brédas, J.-L. Organic photovoltaics. Energy Environ. Sci. 2009, 2, 251–261.

173. Liu, D.; Kelly, T.L. Perovskite solar cells with a planar heterojunction structure prepared using room-temperature solution processing techniques. Nat. Photonics 2013, 8, 133–138.

174. Chen, S.-G.; Stradins, P.; Gregg, B.A. Doping highly ordered organic semiconductors: Experimental results and fits to a self-

consistent model of excitonic processes, doping, and transport. J. Phys. Chem. B 2005, 109, 13451–13460.

175. Pan, S.; Yang, Z.; Chen, P.; Fang, X.; Guan, G.; Zhang, Z.; Deng, J.; Peng, H. Carbon nanostructured fibers as counter electrodes in wire-shaped dye-sensitized solar cells. J. Phys. Chem. C 2014, 118, 16419–16425.

176. Cai, X.; Hou, S.; Wu, H.; Lv, Z.; Fu, Y.; Wang, D.; Zhang, C.; Kafafy, H.; Chu, Z.; Zou, D. All-carbon electrode-based fiber-shaped dye-sensitized solar cells. Phys. Chem. Chem. Phys. 2012, 14, 125–130.

177. Hou, S.; Cai, X.; Fu, Y.; Lv, Z.; Wang, D.; Wu, H.; Zhang, C.; Chu, Z.; Zou, D. Transparent conductive oxide-less, flexible, and highly efficient dye-sensitized solar cells with commercialized carbon fiber as the counter electrode. J. Mater. Chem. 2011, 21, 13776–13779.

178. Cai, F.; Chen, T.; Peng, H. All carbon nanotube fiber electrode-based dye-sensitized photovoltaic wire. J. Mater. Chem. 2012, 22, 14856–14860.

179. Zhang, L.; Shi, E.; Li, Z.; Li, P.; Jia, Y.; Ji, C.; Wei, J.; Wang, K.; Zhu, H.; Wu, D.; et al. Wire-supported CdSe nanowire array photoelectrochemical solar cells. Phys. Chem. Chem. Phys. 2012, 14, 3583–3588.

180. Grätzel, M. The light and shade of perovskite solar cells. Nat. Mater. 2014, 13, 838–842.

181. McGehee, M.D. Perovskite solar cells: Continuing to soar. Nat. Mater. 2014, 13, 845–846.

182. Zhou, H.; Chen, Q.; Li, G.; Luo, S.; Song, T.-b.; Duan, H.-S.; Hong, Z.; You, J.; Liu, Y.; Yang, Y. Interface engineering of highly efficient perovskite solar cells. Science 2014, 345, 542–546.

183. Jeon, N.J.; Noh, J.H.; Yang, W.S.; Kim, Y.C.; Ryu, S.; Seo, J.; Seok, S.I. Compositional engineering of perovskite materials for high-performance solar cells. Nature 2015, 517, 476–480.

184. Qiu, L.; Deng, J.; Lu, X.; Yang, Z.; Peng, H. Integrating perovskite solar cells into a flexible fiber. Angew. Chem. Int. Ed. 2014, 53, 10425–10428.

185. Liu, J.; Mirri, F.; Notarianni, M.; Pasquali, M.; Motta, N. High performance all-carbon thin film supercapacitors. J. Power Sources 2014, 274, 823–830.

186. Wang, G.; Zhang, L.; Zhang, J. A review of electrode materials for electrochemical supercapacitors. Chem Soc Rev 2012, 41, 797–828.

187. Yu, Z.; Tetard, L.; Zhai, L.; Thomas, J. Supercapacitor electrode materials: Nanostructures from 0 to 3 dimensions. Energy Env. Sci 2015, 8, 702–730.

188. Yang, P.; Xiao, X.; Li, Y.; Ding, Y.; Qiang, P.; Tan, X.; Mai, W.; Lin, Z.; Wu, W.; Li, T.; et al. Hydrogenated ZnO core–shell nanocables for flexible supercapacitors and self-powered systems. ACS Nano 2013, 7, 2617–2626.

189. Jiang, H.; Yang, L.; Li, C.; Yan, C.; Lee, P.S.; Ma, J. High–rate electrochemical capacitors from highly graphitic carbon–tipped manganese oxide/mesoporous carbon/manganese oxide hybrid nanowires. Energy Environ. Sci. 2011, 4, 1813–1819.

190. Li, Q.; Wang, Z.-L.; Li, G.-R.; Guo, R.; Ding, L.-X.; Tong, Y.-X. Design and synthesis of MnO2/Mn/MnO2 sandwich-structured nanotube arrays with high supercapacitive performance for electrochemical energy storage. Nano Lett. 2012, 12, 3803–3807.

191. Zhang, Z.; Chen, X.; Chen, P.; Guan, G.; Qiu, L.; Lin, H.; Yang, Z.; Bai, W.; Luo, Y.; Peng, H. Integrated polymer solar cell and

electrochemical supercapacitor in a flexible and stable fiber format. Adv. Mater. 2014, 26, 466–470.

192. Jiang, D.; Jin, Z.; Henderson, D.; Wu, J. Solvent effect on the pore-size dependence of an organic electrolyte supercapacitor. J. Phys. Chem. Lett. 2012, 3, 1727–1731.

193. Jost, K.; Perez, C.R.; McDonough, J.K.; Presser, V.; Heon, M.; Dion, G.; Gogotsi, Y. Carbon coated textiles for flexible energy storage. Energy Environ. Sci. 2011, 4, 5060–5067.

194. Pasta, M.; La Mantia, F.; Hu, L.; Deshazer, H.D.; Cui, Y. Aqueous supercapacitors on conductive cotton. Nano Res. 2010, 3, 452–458.

195. Prosini, P.P.; Pozio, A.; Botti, S.; Ciardi, R. Electrochemical studies of hydrogen evolution, storage and oxidation on carbon nanotube electrodes. J. Power Sources 2003, 118, 265–269.

196. Zou, D.; Lv, Z.; Cai, X.; Hou, S. Macro/microfiber-shaped electronic devices. Nano Energy 2012, 1, 273–281.

CHAPTER 8

Functional Design of the Woven Filters

Lucica Cioară[1] and Ioan Cioară

[1] *Technical University "Gheorghe Asachi"Iassy/Faculty of Textilesand Leather Engineering, România*

1. INTRODUCTION

The filtration process implies the physical separation of one or more components of a fluid that passes through or over a barrier which is permeable to only one or some of the fluid components. Therefore the fundamental element of the filtration process is the barrier which is permeable to only a part of the suspension or the solution applied to filtration. This barrier is named filter medium and the mechanical structure used to support it is named filter. The statement the heart of any filter is the filter medium is fully justified. The most ingenious filter is useless if does not have an adequate filter medium. A specific shape of a filter can use a wide variety of filter mediums to do the same or different separation.

Function of their purpose filtration processes are used to separate solid – gas, solid – liquid, liquid – liquid or solid – solid mixtures. Solid – gas separation domain is represented mainly by air filters also including gas processing. Solid – liquid separation is the usual area of mechanical filters from which a relevant part is the inertial separators. Liquid – liquid and solid – solid separations are complex and specialized areas of filters or separators typology.

Industrial installations frequently use as filtration media technical textiles obtained by weaving i.e. woven filters (Adanur, 1995; Harracks&Anand, 2000). To respond to imposed exigencies by the use in industrial installations, the fabrics utilized as filter media must comply with a wide range of demands which are for the most part determined by the fabrics own structural characteristics and partially by the fabric finishing methods (Marchiş et al.,1991; Cioară et al., 1991; L.Cioară&I.Cioară, 2001). Among these requirements the following are mentioned:

- high filtration capacity, high degree of filtered elements purification and minimum hydraulic resistance;
- good mechanical resistance and stability to chemical, thermal, corrosive and biological agents;
- a high degree of the filtering surface smoothness allowing an easy and total residue separation and filter regeneration;
- a firm and homogenous structure allowing a high filtering process fineness and quality throughout the service life of the filtering element.

The fields having a vital requirement for woven textile media filtration are in a continuous diversification and specialization. It is widely known that technical textiles represent viable alternatives for all economic and social life sectors. In this context the woven textiles filter media, a significant representative of technical textiles, find their application in various fields from automobiles to space industries, in construction, in agriculture and environment protection (Adanur, 1995; Harracks&Anand, 2000).

Filtration processes are accomplished following two basic principles: depth filtration and surface filtration (Medar&Ionescu, 1986). Both forms of the filtering process imply the simultaneous occurrence, in different ratio of two physical phenomena:

- direct particles retention – do to this process filter medium stand for a mechanical barrier for particles bigger than the restriction;
- adsorption – particles retention by electrostatic forces or molecular attraction of the filter medium.

Principle of depth filtration and the specific mechanisms through which the particles are retained are shown in Figure 1.a. Direct interceptions take place when particles of a certain size are passing through larger pores and are trapped in the filter medium structure when meet smaller pores. Inertial forces cause particles to hit filter medium fibers the particles retention being obtained due to their penetration into the body of fibers or due to the fiber deformation.

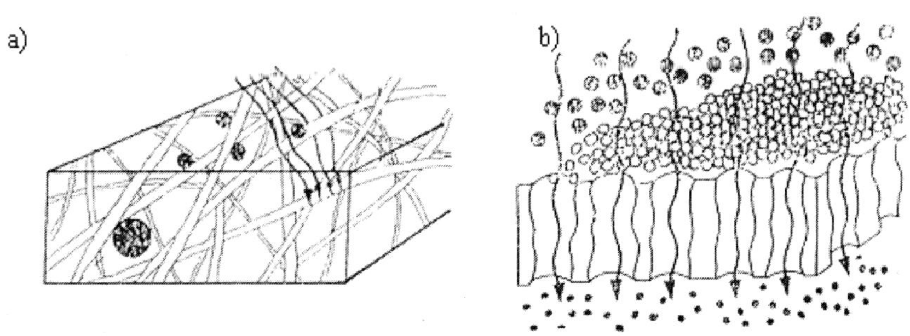

Figure 1. Filtrations mechanisms

Inertial impaction is predominant when high fluid velocity or very dense filter medium is present. This type of filtration mechanism is most predominant when high gas velocity and/or dense packing of the filter media is present. Inertial impaction occurs also when an abrupt change in streamline take place. In this case the particle, due to its inertia, will continue along its original path and could be retained by the filter medium

Adsorption phenomena determine the attraction of small size particles by the filter medium fibers. The adsorption is favored by particles Brownian movement of the particles during the filtration process. Textile filter media that work by the depth filtration principle are: nonwoven fibrous layers, simple textiles made of spun or filamentary yarn, pile or felted, composite fabrics made as semi double, double or multiple layers structures.

Surface filtering implies that particles larger than the pore size are retained on the filter medium surface (Figure 1.b) Due to the adsorption forces particles smaller than pore size can be retained along the pore wall, reducing its transverse dimension causing blocked pores and filter medium clogging as a result. In the first phase of the clogging nominal fineness of filtration is reduced, the pressure difference increases and a combination of surface filtration with a pseudo-depth filtration take place (occurs).

Later on, as the degree of clogging increases, fluid flow through the filter medium is significantly reduced. Textile filter media which operate by surface filtration are monofilament yarns woven textiles.

The comparative analysis of the woven filter media working according to these two principles highlights their advantages and disadvantages (Table 1). In all cases the filter media is considered within the conventional filtering range ensuring the separation of particles over 1 μm in size (Rouette, 2001).

Filtration fineness is influenced by filter medium structure. Pore size distribution is Gaussian for filter media that operates on depth filtering principle and covers a narrower range around the mean value for filter media that works on surface filtration principle. As a result, the particles retention set in the case of surface filtration is much more restrictive than in the case of depth filtration.

Table 1. Comparative analysis of filter media

Depth filtration		Surface filtration	
advantages	disadvantages	advantages	disadvantages
low cost	hard to clean	possibility of cleaning and reuse	high cost
high efficiency	filter medium particles migration is possible	filter medium particles migration is excluded	low initial efficiency
high capacity to retain impurities	filtering performance is depending on the fluid viscosity	fatigue resistance, resistance to temperature or corrosive agents	limited capacity to retain impurities
	relatively large pressure drop	lower pressure drop	
	increased clogging	reduced clogging	

2. ANALYSIS OF WOVEN FILTER MEDIA FUNCTIONALITY

Woven filter media are products that are differentiated by structure and properties in strict accordance with the requirements and particularities of the process in which they operate. The filter medium structure is necessarily associated with the principle used to separate the mixture particles (surface filtration or depth filtration).

2.1. Features woven filter media

The result of filter medium different properties combination sets up its quality and respectively its functionality. For an objective assessment of filter media quality (functionality) three groups of properties have been identified as follows:

- properties related to filter medium mounting system type. Those properties are important for the mechanical implementation of the filter respectively the filter medium set up on the support frame.

Among the key properties of this group stated: stiffness, tensile strength, tear resistance, burst strength, abrasion resistance, vibration stability, elongation, the edges stability;

- properties related to the application type that are taking in consideration the compatibility between the filter medium and the processed medium. In this category falls the following properties: chemical stability, thermal stability, biological stability, dynamic stability, adsorption, absorption, operational safety and security, electrostatic characteristics, reuse capability, price;
- properties addressing specific filtrations process particularities underlining the filter medium capacity to comply with required demands. The most important properties of this group are: the smallest particle retained, retention efficiency, the structure of filter media, particle shape, filtering mechanisms used, flow resistance, porosity of filter media, permeability, tendency to clog, filter-cake discharge characteristics.

For each filter medium, depending on field of use and the requirements in service only some of these properties are necessary. As a result, the design of woven textiles intended to be used as filter media must be made in accordance with functionality criteria ensuring priority to the properties requested by the process utilized.

The relation structure – properties – use value is the design criterion for woven filters. Value engineering is a method of research and systemic design according to which the functions of the product studied (filter medium) must be designed and carried out with minimum expenditure in terms of highest quality, reliability and performance (Condurache et al., 2004). Value engineering instrumentation methodology implies the following stages:

- functional analysis: answers the questions what is and what the product does; the function list of the analyzed product is completed;

- classification of functions: answers the question how important the function is and how well meets the user requirements; function's relative importance, intrinsic and technical dimension terms are ascertained; functions classification for the analyzed product is finalized;
- product design or redesign based on required functions.

Table 2. The functions of woven filter media

Symbol	Function name	Technical dimension	Function type
F1	to separate the phases of a heterogeneous mixture	porosity permeability	primary, objective, necessary, general
F2	to ensure filtration fineness	shape, size and pore distribution	primary, objective, necessary, specific to filter principles
F3	to be dimensionally stable during operation	elastic and residual tensile strain	primary, objective, necessary, specific to filter principles
F4	to withstand the action of mechanical factors during operation	tensile strength burst resistance	primary, objective, necessary, specific to filtration process
F5	to withstand the erosive effects of the environment	chemical resistance	primary, objective, necessary, specific to filtered fluid
F6	to ensure filtration velocity	active filtration surface	primary, objective, necessary, specific to filter principle
F7	to withstand the erosive action of the filtered fluid	abrasion resistance	secondary, objective, necessary, specific to filtration process
F8	mechanical durability	fatigue resistance	secondary, objective, necessary, specific to filter type
F9	shouldn't clog	filter structure, pores shape	secondary, objective, necessary, specific to filter type
F10	easy to clean and rebuild	filter structure raw material	secondary, objective, necessary, specific to filter type
F11	easy to fabricate	filter structure, fabrication technology	secondary, objective, necessary, specific to filter type
F12	easy to install and replace	filter shape and dimensions	secondary, objective, necessary, specific to filter type

- Function is considered an essential attribute of the studied product expressed in terms of medium and user. In the same time, the function can be regarded as a characteristic of the product that determines a particular utility. The list of function classification is the starting point of value engineering studies.
- Product functions are determined by importance, the measurement method, its contribution to achieving use value and the degree of generality.
- Drawing up the list of filter medium functions is based on defining the filter medium and the conditions under which it works. In principle woven filters are intended to be used in filtration.

Based on such considerations the functions considered to be necessary for the filter medium, their technical elements of assessment and their typology classified by standard criteria are shown in Table 2 (I.Cioară&L.Cioară, 2009).

2.2. The hierarchy functions woven fabrics filter

After developing the list of all functions their classification is done in order to establish the importance and weightiness of each function in rapport to all functions the product offers. The classification of the function is done using Value Engineering methods such as the Expertise Method or Imposed Decision Method. Regardless of the method used the classification is done considering all filter media functions or, selectively, group of functions, classified according to their typology.

Imposed Decision Method presents a high degree of objectivity (Condurache et al., 2004). To apply this method the following steps must be achieved: comparing the functions in pairs, calculating the importance coefficient for each function and classifying the functions by their importance coefficient value. By comparing the environmental functions as a filter to obtain decisions (0-1), (0.5-0.5) or (1-0). Scoring are considered: 0 considered less important function, 1 for the function considered more important, compared to 0.5 when the functions are valued as important. D total number of decisions resulting from the comparison of the n features of the filter is calculated with:

$$D = C_n^2 = \frac{n \cdot (n-1)}{2} \qquad (1)$$

I importance factor for each sample is calculated function the relationship:

$$I = \frac{N}{D} \qquad (2)$$

where: N is the sum of points awarded;

D - total number of decisions.

The filter media 12 functions obtained by weaving defined in Table 2 were divided into two groups: 6 primary and 6 secondary functions. Apply for group relationship of the main functions, which will be used to design, to establish the number of decisions as follows:

$$D = C_6^2 = \frac{6 \cdot (6-1)}{2} = 15 \qquad (3)$$

Table 3. Coefficient calculation Ranking

Function	Decisions															N	I
	1	2	3	4	5	6	7	8	9	10	11	12	13	14	15		
F1	0.5	0.5	1	1	0.5											3.5	0.233
F2	0.5					0.5	1	1	0.5							3.5	0.233
F3		0.5				0.5				0.5	0.5	0.5				2.5	0.166
F4			0				0			0.5			0.5	0		1	0.066
F5				0				0			0.5		0.5		0	1	0.066
F6					0.5				0.5			0.5		1	1	3.5	0.233

Table 4. Priority functions of the filter media and their assessment criteria

The name of the function	Specific technical dimension	Structural characteristics of fabric
to separate the phases of a heterogeneous mixture	pore size	fineness and density of yarns
to ensure filtration fineness	pore shape and distribution, filter medium fineness	fineness and density of yarns, weave
to ensure filtering velocity	adequate filtering active area	fineness and density of yarns
to be dimensionally stable during operation	structural and mechanical characteristics of yarn and fabric	the mechanical characteristics of the of yarns
to withstand the action of mechanical factors during operation	tensile strength burst resistance	the mechanical characteristics of the of yarns
to withstand the erosive effects of the environment	chemical resistance	the nature of raw material

In Table 3 are comparative analysis, two by two principal functions. The last column of the table are shown the importance scores and values of each corresponding functions. Based on the values of the coefficients of importance to obtain the hierarchy of the main functions in the following sequence: F1, F2, F6, F3, F4, F5.

Priority functions set out in this way are taken into consideration when designing or redesigning filter media weaved in accordance with functionality criteria. In this respect the assessing criteria of woven filter media priority functions are summarized in Table 4.

The parameters specific to the woven filter with simple structure that will provide functional design criteria are: the relative porosity, pore shape and size, the pore distribution, the active filtering surface.

The structural characteristics of woven fabric,which determine the parameters of filter fabric are: yarns count, thread density and weave.

3. STRUCTURAL AND FUNCTIONAL CHARACTERISTICS OF THE FABRIC FILTERS WITH SIMPLE STRUCTURE

Characterization and use of the fabric estimating filters with simple structure can be made by means of specific structural and functional characteristics (Behera, 2010; Cioară, 2002). Among these characteristics are mentioned: porosity, pore size and architecture, active filtering surface environment and filter fineness.

3.1. Porosity

Porosity, feature size filter material is the property of having pores in their structure (Medar&Ionescu, 1986; Cay et al., 2005). In connection with the porosity are two distinct notions:

- relative porosity, apparent or open, when taking into account only pores that communicate with each other;
- absolute porosity, effective or real, if we take into account all the pores, i.e. those who are isolated.
- Fluid flow through uniform or uneven spaces created by the filter medium, while maintaining the quality of filtration, filtration efficiency and smoothness and filtering capacity are issues directly related to the porosity of filter media.
- Fluid movement across the filter medium is described by the filtration rate, defined as the maximum volume of fluid passing per unit time through unit area of filter. Porosity refers to the filter media pore volume per unit volume and is typically seen in relative units. Generally, the textile filtering media are inhomogeneous because the filter permeability changes during the exploitation. The medium in homogeneity can be bigger or smaller, depending on the structure of woven filter.

3.2. Pore dimensions and architecture

An important feature of each filter surfaces is the existence of pores which penetrate the entire thickness of the filter and retain solid particles larger than the pores in the cross section of their most narrow, but allow passage of fluid that carried them. Small pore is a void within a solid body. After dimensions are distinguished (Medar&Ionescu, 1986) : fine pores with a diameter greater than 20 μm (invisible to the naked eye) and coarse pore diameter greater than 20 μm (visible to the naked eye). The way of communication with the outside pores can be:

- open, when communication with the outside;
- closed, when no communication with the outside.

Dimensional uniformity and stability of pores of a filter medium directly influences the process of filtering performance (Gabrijelcic et al., 2009). Pore size and shape of woven textile filter media are dependent on the basic structural parameters of fabric: the fineness of yarns, thread density and the weave.

Pore's characteristics which is assessed functional performance of a fabric filter are: side pore, pore area, architecture and distribution of pores in the fabric plane.

Side and pore area are geometric features of woven fabric due to its basic structural parameters.

In terms of basic structural parameters, the woven fabrics with simple structure can be balanced or unbalanced in fineness and density yarns, and the resulting pores have square or rectangular form.

In Figure 2 are defined pore geometry of a woven fabric with simple structure. In the balanced woven fabric (Figure 2.a), the warp and weft, have the same fineness and density, the same diameter d and the same density P. Therefore to obtain a filter fabric with square pores. In the unbalanced woven fabric (Figure 2.b), the warp and weft have different count, expressed by d_u, d_b, and different density threads expressed by Pu, Pb. As a consequence to obtain a fabric filter with rectangular pores.

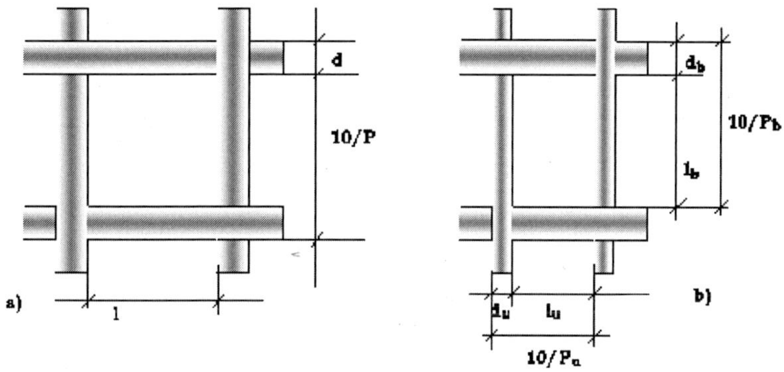

Figure 2. Pore geometry

3.2.1. Pore side

Pore side is the distance between two consecutive threads of the fabric measured in the projection on the horizontal fabric's plan.

For structures balanced (Figure 2.a), pore side it could be calculated with:

$$l = \frac{10}{P} - d \ \text{(mm)} \tag{4}$$

For structures unbalanced (Figure 2.b), pore sides, l_u and l_b, is defined by relations:

$$l_u = \frac{10}{P_u} - d_u \ \text{(mm)}; \ l_b = \frac{10}{P_b} - d_b \ \text{(mm)} \tag{5}$$

3.2.2. Pore area

Pore area is defined as the projection on the horizontal plan of the fabric's pore.

For balanced structure (Figure 2.a) the pore area A_p is calculated with:

$$A_p = l^2 \text{ (mm}^2) \quad (6)$$

For unbalanced structure (Figure 2.b) pore area is calculated by the relationship:

$$A_p = l_u \cdot l_b \text{ (mm}^2) \quad (7)$$

3.2.3. Pores architecture

Pores architecture is a characteristic determined by the fineness of yarns, threads density, mechanical and rheological characteristics of yarns and weave used. The segments of the yarn which constitute the pore sides can be considered, like a beam in one of two situations:

- passing from one side to another fabric,
- above or under the opposite yarn system.

Under these circumstance, the pores` shape and size depend not only on yarn count and density, but also on the positions of the yarns in the weave.

For this purpose an analysis is presented which highlights the fact that the weave determines the distribution of requests in the yarns and the default form of pores. The analysis is done on three woven fabrics filter, whose characteristics are presented in Table 5.

The weave used in implementing the three fabrics are shown in Figure 3. To the right of each weave are represented by pores with a distinct architectural (Cioara et al., 2003).

Table 5. Variants of filter fabrics

Woven filter	Weave	Raw material	Yarn diameter (mm)	Thread density (yarns/cm)
Filter 50 mesh	Plain	Polyamide	0.14	20
Filter 22 mesh	Twill D2/2	Polyamide	0.45	9
Filter 24 mesh	Twill D3/1	Polyamide	0.45	9.5

After examining the shape of pores in the three weave can be made the following observations and interpretations:

- at the fabric filter with plain weave (Figure 3.a) all the pores have the same architecture; the threads have a similar position in the pore sides (all threads are crossing from one side to another of the fabric). Under these conditions the fabric structure creates the potential formation of uniform pores in the shape of their; the pores I, is identical in structure with pores II;

- at the fabric filter with twill weave D 2/2 (Figure 3.b) pores of the report have the same structure. Shapes and sizes of the four types of pores are identical. It creates the conditions to achieve a uniform structure with a high degree of homogeneity to ensure a quality filter; pores numbered I, II III IV is identical in structure;

- at the fabric filter with twill weave D 1/3 (Figure 3.c) is classified in four types of pores. The four distinct architecture creates pores with different shapes and volumes that the conditions for the flow through it are differentiated; pores numbered II, is identical in structure with pores IV, pores I and III is different.

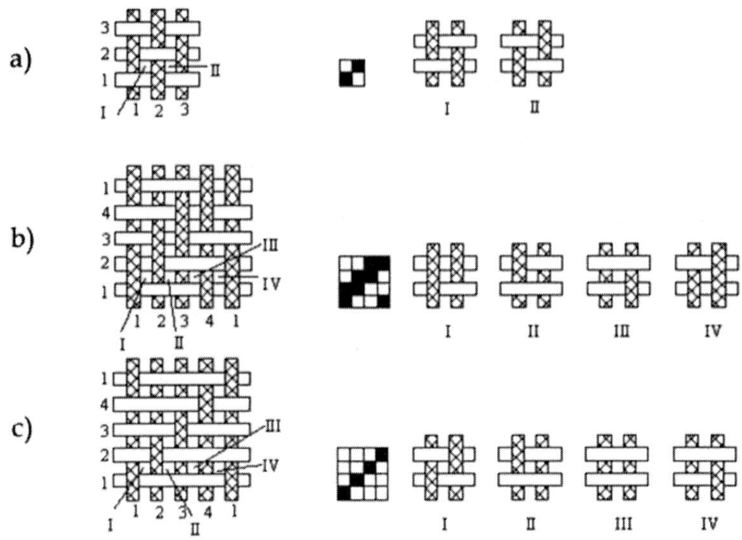

Figure 3. Type of pores

To support these interpretations, on the three types of fabric filter were performed measurements of area pores, using a specialized program. Measurements were made on the fabric images captured with a stereo microscope with digital camera.

These images are presented in Figure 4, 5 and 6. In each figure is shown a fragment of the microscopic image of the fabric filter with the contours measured pore, the statistics of individual value of area and pore distribution curve.

To facilitate analysis of information provided by research, in Table 6 were centralized statistical processing on the string values of individual values of the areas measured.

The analysis of microscopic images shows that the filter fabric with plain weave (Figure 4) pores are relatively uniform shapes and sizes. This is supported on the one hand, the low dispersion of individual values (s=0.00644) and, on the other hand, the restricted distribution of individual values around the average. Extreme values, minimum and maximum, with reduced weight, have a deviation of up to 10% of the average pore area.

Table 6. Statistical evidence

Weave	Mean area (mm^2)	Min. value (mm^2)	Max. value (mm^2)	Range (mm)	Std.Dev. (mm)	CV (%)
Plain	0.1021164	0.0841335	0.1120637	0.0279302	0.0064476	6.31
Twill D2/2	0.2104919	0.1555840	0.2731934	0.1176094	0.0310180	14.73
Twill D3/1	0.3233822	0.1810607	0.4755648	0.2945041	0.0722753	22,34

At the fabric filter with twill weave D 2/2 (Figure 5) is observed as architecture, two types of pores with greater irregularity than plain weave. Pore area shows a greater variation, which is confirmed by the dispersion value (s=0.03101) and the pore distribution curve shape. Even if the pore area varies widely, up to 25% from the mean, the woven filter is estimated that the structure is uniform. Extreme values are numerous and, consequently, the distribution curve is wider.

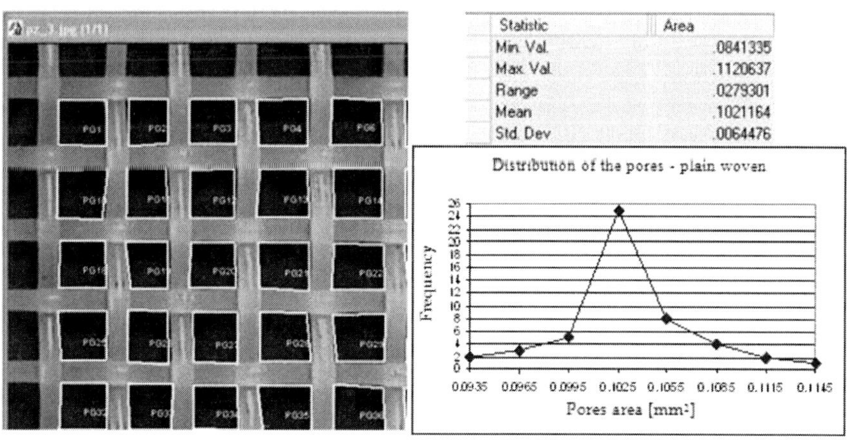

Figure 4. Pore architecture of the woven filter - plain weave

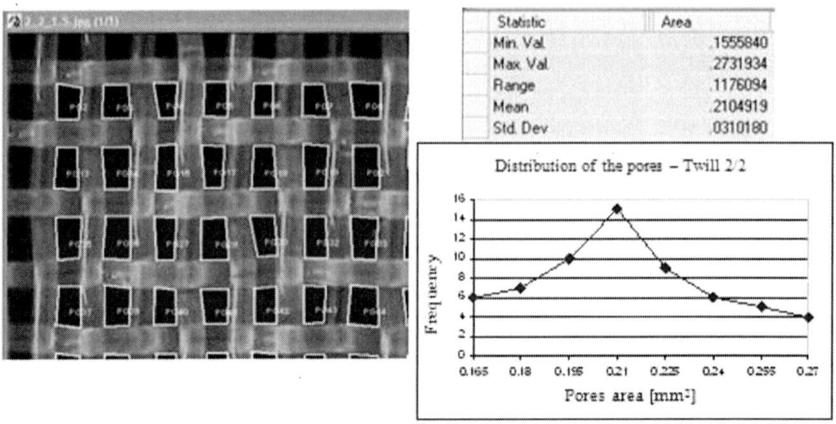

Figure 5. Pore architecture of the woven filter - twill D 2/2

At the fabric filter with twill weave D 3/1 (Figure 6) is observed as architecture, the four specific types of pores. The dispersion of measured values (s=0.07227) indicates large variation in pore area, which is emphasized by the distribution curve. Pore area varies widely, with over 50% of the mean value and the structure of a fabric filter has emphasized the uneven character.

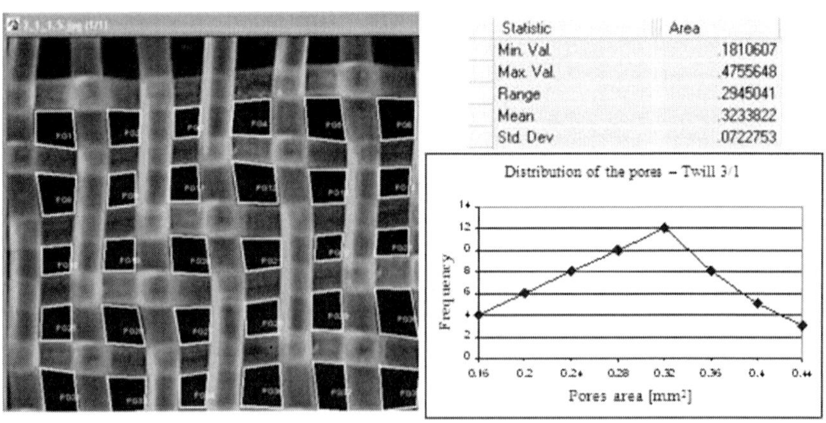

Figure 6. Pore architecture of the woven filter - twill D 3/1

3.3. The active filtering surface

This parameter provides information about the porosity of the woven filter and resistance to fluid flow through it. Active filtering area was defined as the ratio of pore area, A_p, and area of fabric element, A_{et}, is calculated from the relationship:

$$S_a = \frac{A_p}{A_{et}} \cdot 100\ (\%) \tag{8}$$

3.4. Filter fineness

The woven filter fineness is a nominal identification, which is expressed by the number of pores per unit length or number of pores per unit area. The filter fineness is calculated by the following relations:

balanced woven filter fabrics:

$$F = P\ (\text{pores / cm}) \tag{9}$$

$$F_m = 2.54\ P\ (\text{pores / inch}) \tag{10}$$

$$F_d = P^2\ (\text{pores / cm}^2) \tag{11}$$

unbalanced woven filter fabrics:

$$F_d = P_u P_b\ (\text{pores / cm}^2) \tag{12}$$

4. ALGORITHMS FOR FUNCTIONAL DESIGN OF FILTER. EXAMPLES OF APPLICATION

The last stage of the value engineering technique is to design or redesign based priority functions. Based on the conclusions presented in the above analysis were developed five functional design algorithms for simple structure filter fabrics balanced and unbalanced in fineness and/or density (I.Cioara & L.Cioara, 2010). Always choose the algorithm is done according to known elements (input data) and the requirements process of filtering (output data) to design, redesign or verification. Following the sequence of calculation steps is different from one algorithm to another.

4.1. Algorithms for functional design of filter

In Tables 7 and 8 shows the deployment of the calculations according to data input and output, these calculations are completed to the mass of fabric filter.

In the relationships of calculating the parameters of qualitatively appraising the simple structure woven fabrics used as filtering media the significance of the employed symbols is the following: d - yarn diameter in the balanced structures (mm); d_u, d_b - warp and weft diameter (mm); A - tabled constant for the diameter calculating; A_p - pore area, (mm^2); A_{et} - woven fabric element area, (mm^2); T_{tex} - yarn count in the balanced structures, (g/km); T_{texu}, T_{texb} - warp and weft count, (g/km) ; P - threads density in the balanced structures, (yarns/cm); P_u, P_b - warp density and weft density, (yarns/cm); l - square pore side, (mm); l_u, l_b - the pore side in the warp and weft direction, (mm); F - balanced structure filter fineness, (pores/cm); F_m - balanced structure filter fineness, (pores/inch); F_d - filter fineness, (pores/cm^2); M - woven fabric mass, (g/m^2); a - crimp yarn in the woven fabric, (%); m – factor of unbalanced for threads density.

4.1.1. Algorithm I. Design of simple filters based on the density

In this case we consider as known the basic structural characteristics of fabric: warp and weft yarn diameters (d_u, d_b) and their densities (P_u, P_b). The algorithm is used to identify specific characteristics of woven fabrics filters: the filter fineness, the pore side and area, the active filtering surface.

Based on of factors identified can be appreciated if appropriate filter fabric filtration process characteristics. Depending on the conclusions formulated after verification can be redesign the fabric to match the outlays.

4.1.2. Algorithm II. Design of simple filters based on the yarn diameter and filter fineness (Table 7)

In this case it is considered known diameters of the warp and weft of yarns (d_u, d_b) and fineness filter fabrics. The algorithm is used to design a filter of woven fabrics density unbalance. The yarns density, the pores sides and surface, the active filter area and the fabric weight are determined by calculations.

Since by changing the density unbalance factor the pores shape and dimensions are also modified ensuring an appropriate filtering fineness. The algorithm can be additionally used to re-design according to given requirements.

4.1.3. Algorithm III. Design of simple filters based on the yarn diameter and side of pore (Table 7)

The algorithm is used to design the fabric characteristics in compliance with the parameters of the filtering process. The input data are warp and weft yarn diameters (d_u, d_b) and pore sizes (l_u, l_b). Imposing pore side means the particle to be retained is known, the filter being a calibrated restriction. The required yarn density and all the specific characteristics of the filter are obtained by calculations.

Table 7. Specific elements for algorithms design I, II, III

1. Input data: d_u, d_b, P_u, P_b	1. Input data: d_u, d_b, F_d	1. Input data: d_u, d_b, l_u, l_b
2. Yarn count $T_{tex} = \frac{d^2}{A^2}$	2. Yarn count $T_{tex} = \frac{d^2}{A^2}$	2. Yarn count $T_{tex} = \frac{d^2}{A^2}$
3. Filter fineness balanced structures $F = P_u = P_b$ pores / cm $F_m = F \cdot 2.54$ pores/inch unbalanced structures $F_d = P_u \cdot P_b$ pores/cm^2	3. Thread density $P = f(F_d)$ $P_u \cdot P_b = F_d$ $\frac{P_u}{P_b} = m$ $P_u = \sqrt{F_d \cdot m}$ $P_b = \frac{P_u}{m}$	3. Thread density $P = f(l_p, d)$ $P_u = \frac{10}{l_u + d_u}$ $P_b = \frac{10}{l_b + d_b}$
4. Pore side $l_u = \frac{10}{P_u} - d_u$ $l_b = \frac{10}{P_b} - d_b$	4. Pore side $l = f(P, d)$ $l_u = \frac{10}{P_u} - d_u$ $l_b = \frac{10}{P_b} - d_b$	4. Filter fineness balanced structures $F = P_u = P_b$ pores / cm $F_m = F \cdot 2.54$ pores/inch unbalanced structures $F_d = P_u \cdot P_b$ pores/cm^2
5. Pore area $A_p = f(P, d)$ $A_p = \left(\frac{10}{P_u} - d_u\right) \cdot \left(\frac{10}{P_b} - d_b\right)$		
6. Filtering active surface $S_a = f(P, d)$ $S_a = (10 - P_u \cdot d_u) \cdot (10 - P_b \cdot d_b)$		
7. Woven fabric mass $M = \frac{P_u \cdot T_{texu}}{10} \cdot \frac{100}{100 - a} + \frac{P_b \cdot T_{texb}}{10} \cdot \frac{100}{100 - a}$		

4.1.4. Algorithm IV. Design of simple filters based on the yarn diameter and active filter area (Table 8)

Active filter area is a feature that determines the permeability of the fabric and filter potential to withstand the fluid flow. This algorithm is considered known following elements: the warp and weft yarn diameters (d_u, d_b) and active filtering surface (S_a).

Table 8. Specific elements for algorithms design IV, V

Algorithm IV	Algorithm V
1. Input data: d_u, d_b, S_a	1. Input data: l_u, l_b, S_a
2. Yarns count $T_{tex} = \frac{d^2}{A^2}$	2. Pore area $A_p = l_u \cdot l_b$
3. Thread density for balanced structures of cover factory: $P_u \cdot d_u = P_b \cdot d_b$ $P = f(S_a, d)$ $P_u = \frac{10 - \sqrt{S_a}}{d_u}$ $P_b = \frac{10 - \sqrt{S_a}}{d_b}$	3. Thread density $P = f(S_a, l_p)$ $S_a = \frac{A_p}{A_{et}} \cdot 100 \Rightarrow A_{et} = \frac{A_p \cdot 100}{S_a} = \frac{10}{P_u} \cdot \frac{10}{P_b}$ $P_u \neq P_b$; $\frac{P_u}{P_b} = m$ $P_b = \sqrt{\frac{S_a}{m \cdot A_p}}$ $P_u = P_b \cdot m$
4. Filter fineness balanced structures $F = P_u = P_b$ pores / cm $F_m = F \cdot 2.54$ pores/inch unbalanced structures $F_d = P_u \cdot P_b$ pores/cm^2	4. Filter fineness balanced structures $F = P_u = P_b$ pores / cm $F_m = F \cdot 2.54$ pores /inch unbalanced structures $F_d = P_u \cdot P_b$ pores/cm^2
5. Pore side $l = f(P, d)$ $l_u = \frac{10}{P_u} - d_u$ $l_b = \frac{10}{P_b} - d_b$	5. Yarns diameter $d_u = \frac{10}{P_u} - l_u$ $d_b = \frac{10}{P_b} - l_b$
6. Pore area $A_p = f(P, d)$ $A_p = \left(\frac{10}{P_u} - d_u\right) \cdot \left(\frac{10}{P_b} - d_b\right)$	6. Yarns count $T_{tex} = \frac{d^2}{A^2}$
7. Woven fabric mass $M = \frac{P_u \cdot T_{texu}}{10} \cdot \frac{100}{100 - a} + \frac{P_b \cdot T_{texb}}{10} \cdot \frac{100}{100 - a}$	

The algorithm allows the calculation of specific characteristics of the filter: pore area and pore side, which determines the filtration fineness and efficiency.

4.1.5. Algorithm V. Design of simple filters based on the pore side and active filter area (Table 8)

In this case filter fabrics design is based on filtering process requirements. Input data are: pore sizes (l_u, l_b) and active filtering surface (S_a). Active filter area determines the filter permeability and influences the flow velocity. The pore side determines the size of the retained particle and influences the filtration fineness and efficiency. The structural characteristics of the fabric are obtained by calculations.

4.2. Examples of application

Further gives examples of the implementation of the five algorithms to design filter woven fabrics with simple structure made from monofilament yarns.

To facilitate the calculation has been made a computer assisted design program (Cioara et al., 2008).

Sequence with the principal program menu is shown in Figure 7. At the top of the screen, interactive buttons, the corresponding input data are highlighted five algorithms. The dates presented in Figure 7 are typical application example of the algorithm I (Table 9).

Table 9. Balanced structure - designed with algorithm I (weave plain)

Sample filter	Parameter	Symbol	Value	UM
	Yarns diameter	d	**0.31**	mm
	Thread density	P	**11.0**	threads/cm
	Filter finess	F	11.0	pores/cm
	Filter finess mesh	F_m	27.94	pores/inch
	Pore side	l	0.599	mm
	Pore area	A_p	0.3589	mm²
	Filtering active surface	S_a	43.43	%
	Dimensional factor	A	**0.0392**	-
	Yarns count	T	62.54	tex
	Crimp yarn	a	4	%
	Woven fabric mass	M	143.32	g/m²

The program has the possibility of saving data Input and account for the different variants. These data are provided to the operator in the form of tables. In this way the designer can to compare specific elements of the different variants and select the variant that best corresponds to requirements process of filtering.

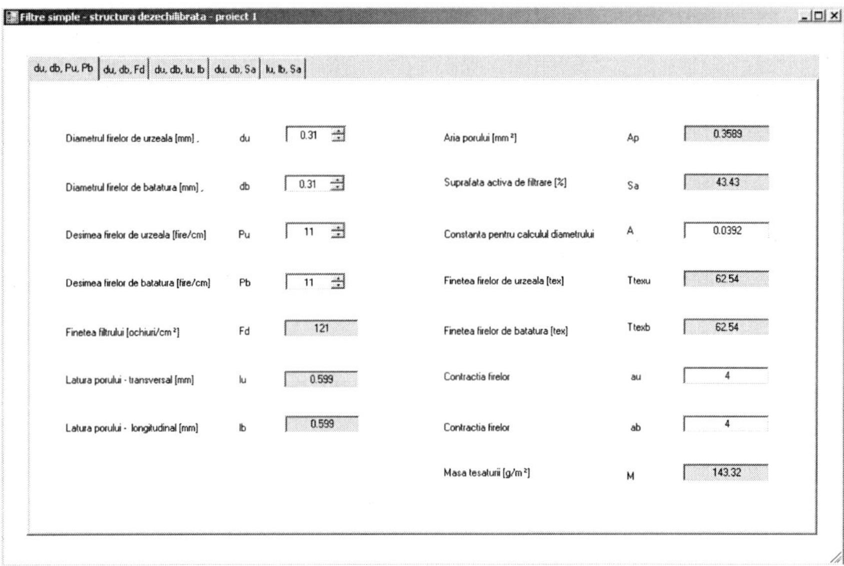

Figure 7. The main menu of the program

Table 10. Balanced structure - designed with algorithm II (weave – hopsack 2/2)

Sample filter	Parameter	Symbol	Value	UM
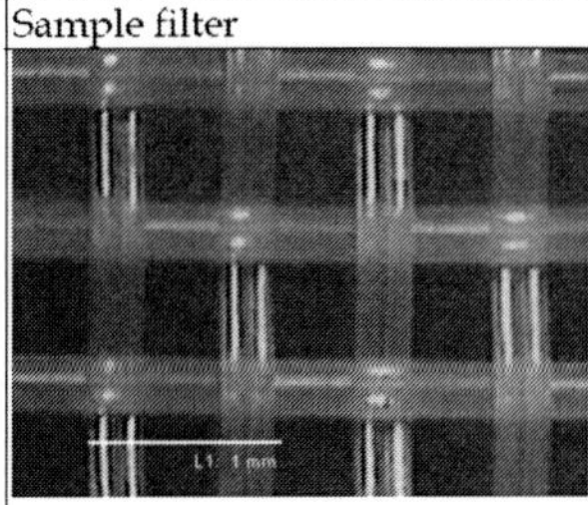	Yarns diameter	d	0.29	mm
	Filter finess mesh	Fm	36	pores/inch
	Filter finess	F	14.2	pores/cm
	Thread density	P	14.2	threads/cm
	Pore side	l	0.414	mm
	Pore area	A_p	0.1716	mm^2
	Filtering active surface	S_a	34.6	%
	Dimensional factor	A	0.0392	-
	Yarns count	T	54.73	tex
	Crimp yarn	a	2	%
	Woven fabric mass	M	158.61	g/m^2

The applications are presented in Tables 9, 10, 11, 12, 13. These applications are made for woven fabrics filters with finesse and densities balanced structures, made with weave plain, weave – hopsack 2/2, weave-irregular sateen and weave – twill 2/2.

Table 11. Balanced structure - designed with algorithm III (weave-irregular sateen)

Sample filter	Parameter	Symbol	Value	UM
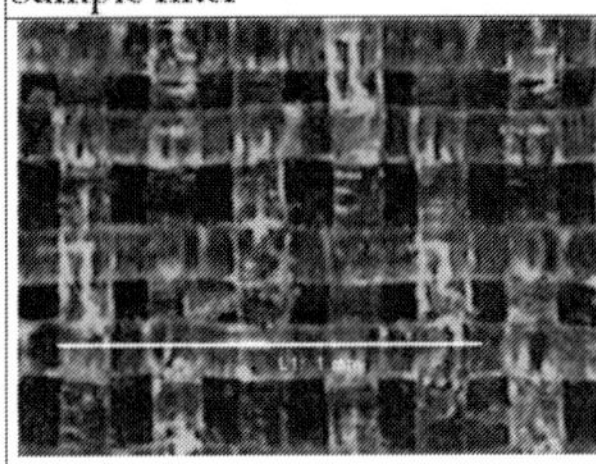	Yarns diameter	d	**0.11**	mm
	Pore side	l	**0.11**	mm
	Thread density	P	45.5	threads/cm
	Filter finess	F	45.5	pores/cm
	Filter finess mesh	Fm	115.57	pores/inch
	Pore area	Ap	0.0121	mm^2
	Filtering active surface	Sa	24.95	%
	Dimensional factor	A	**0.0392**	-
	Yarns count	T	7.87	tex
	Crimp yarn	a	4	%
	Woven fabric mass	M	74.6	g/m^2

Table 12. Balanced structure - designed with algorithm IV (weave – plain)

Sample filter	Parameter	Symbol	Value	UM
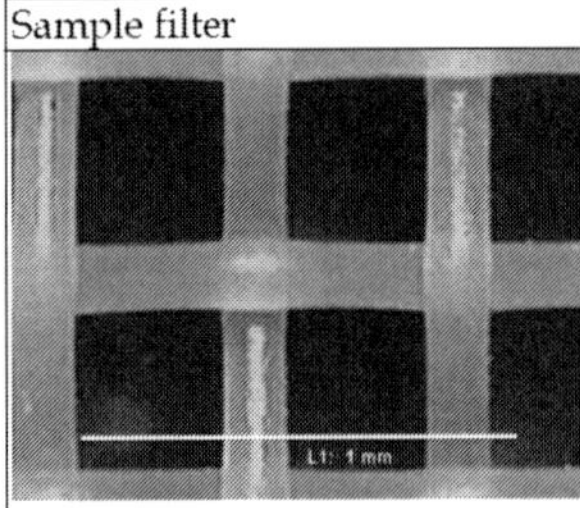	Yarns diameter	d	**0.14**	mm
	Filtering active surface	S_a	**49**	%
	Thread density	P	21.4	threads/cm
	Filter finess	F	21.4	pores/cm
	Filter finess mesh	F_m	54	pores/inch
	Pore side	l	0.327	mm
	Pore area	A_p	0.107	mm^2
	Dimensional factor	A	**0.0392**	-
	Yarns count	T	12.76	tex
	Crimp yarn	a	5	%
	Woven fabric mass	M	57.49	g/m^2

Each table is presented as a fabric filter microscopic image and the list of parameters supplied by the computer program. Same time, the tables are highlighted input data specific to each algorithm separately.

Table 13. Balanced structure - designed with algorithm V (weave – twill 2/2)

Sample filter	Parameter	Symbol	Value	UM
	Pore side	l	0.1	mm
	Filtering active surface	S_a	31	%
	Pore area	A_p	0.01	mm²
	Thread density	P	55.7	threads/cm
	Filter finess	F	55.7	pores/cm
	Filter finess mesh	F_m	141.4	pores/inch
	Yarns diameter	d	0.08	mm
	Dimensional factor	A	0.0392	-
	Yarns count	T	4.16	tex
	Crimp yarn	a	6	%
	Woven fabric mass	M	49.3	g/m²

In each example there is a good correlation between the elements provided by the computer program and structural characteristics shown on the microscopic image of the fabric made. This aspect allows us to say that the proposed algorithms achieve a better modeling of the structural parameters and specific functional of woven fabrics filters with simple structure.

5. CONCLUSION

The woven fabrics which are used as filter fabrics, have the functionality imposed for the filtration process. Structure and properties of the woven filter fabrics are adequately differentiated, for the principles of the filtering process. The paper defines the structural and functional elements that are specific to the filtering woven fabrics, which have a structure that is simple, balanced and unbalanced in yarn count and thread density.

The methods of filter design for the fabrics with simple structure is based on the specific geometry of the structure elements fabrics. To achieve filter fabric with uniform pore (size and shape), is recommended of use, weaves balanced with equal segments and uniform distribution (plain, twill D 2/2, hopsack 2/2). Five algorithms of designing the filtering woven fabrics having a simple structure were elaborated, which can be differentially applied depending on the initially introduced elements.

By using computer aided design program to obtain a set of variations of the designer select the optimal variant, which corresponds better to the needs of field exploitation.

6. ACKNOWLEDGEMENTS

The paper was developed in the research program PNCDI II no.31-088/2007 contract, financed by the Romanian government, which the Technical University "Gheorghe Asachi" of Iassy was a partner. Research presented in this work were performed at the Faculty of Textiles and Leather Engineering of Iassy, Department Technology and Design of Textile Products, laboratory TEXTILEXPERT.

REFERENCES

1. S. Adanur, (1995, (1995).Handbook of Industrial Textiles, Technomic Publishing Company, 1-56676-340-1 U.S.A.
2. Behera, B.K.&Hari,P.K. (2010).Woven textile structure, Woodhead Publishing Limited, ISBN 978-1-84569-514-9, Cambridge, UK
3. A. Cay, S. Vassiliadis, M. Rangoussi, I. Tarakcioglu, (2005, (2005).On the use of image processing techniques for the estimation of the porosity of textile fabrics, World Academy of Science, Engineering and Technology, 2 2 76 79 , 2010-3778
4. I. Cioară, L. Cioară, (2010, (2010).Algorithms design of the filtering woven fabrics with simple structure, Annals of the University of Oradea- fascicle of Textiles Leatherwork, volume XI, 41 46 , 1582-5590 Oradea, Romania

5. I. Cioară, L. Cioară, (2009, (2009).Criteria for estimating the functionality of the filtering mediaobtained by weaving, Industria Textilă, 1 21 25 , 1222-5347 Bucharest

6. I. Cioară, L. Cioară, D. Cascaval, (2008, (2008).Method and software design woven filter media, The VI National Conference of Virtual Learning (CNIV), University of Bucharest pp.60 6 . Publishing, I. S. S. N. 1842-470, Constanţa, România

7. I. Cioară, L. Cioară, E. Onofrei, 2003 Aspects regarding the architecture of the filtering woven fabrics, UNITEX, 1 28 30 , Belgium

8. L. Cioară, (2002, fabric. Woven, Performantica. structure, House. I. S. B. N. Publishing, 973-8075-1, 1. , Iassy, România

9. L. Cioară, I. Cioară, 2001 Optimization of the technical parameters in processes of filter weaving, Proceedings vol.I, 1st Autex Conference Technitex-Technical Textiles: Designing Textiles for Technical Applications, 9-72984-683-9 467 470 , Minho

10. L. Cioară, I. Cioară, O. Marchiş, 1991 Thc influcncc of ccrtain technical parameters onthe capacity of filtering the woven textile planes, Industria Uşoară, 4 239 241 , 1222-5347 Bucharest, România

11. G. Condurache, R. Ciobanu, M. Niculae, (2004, (2004).Analysis and value engineering, Performantica Publishing House, 973-730-022-x Iassy, România

12. H. Gabrijelcic, R. Urbas, F. Sluga, K. Dimitrovski, (2009, (2009).Influence of Fabric Constructional Parameters and Thread Colour on UV Radiation Pretection, FIBRES&TEXTILES in Eastern Europe, 17 17 1 72), 46 54 , 1230-3666 Lodz, Poland

13. A. R. Harracks, S. C. Anand, 2000 Handbook of technical textiles, Woodhead Publishing Limited, 1-85573-385-4

14. O. Marchiş, L. Cioară, I. Cioară, (1991, (1991).Considerations on certain factors that influence the porosity of filtering planes, Industria Uşoară, 4 4 237 238 , 1222-5347 Bucharest, România

15. S. Medar, F. Ionescu, 1986 Filters for hydraulic and pneumatic drives, Technical Publishing House, Bucharest, România

16. H. K. Rouette, 2001 Encyclopedia of Textile Finishing, Woodhead Publishing Limited, 1-84569-065-6

CHAPTER 9

Medical Smart Textiles Based on Fiber Optic Technology: An Overview

Carlo Massaroni, Paola Saccomandi and Emiliano Schena *

Center for Integrated Research, Università campus Bio-Medico, Alvaro del Portillo, 21, Rome 00128, Italy

ABSTRACT

The growing interest in the development of smart textiles for medical applications is driven by the aim to increase the mobility of patients who need a continuous monitoring of such physiological parameters. At the same time, the use of fiber optic sensors (FOSs) is gaining large acceptance as an alternative to traditional electrical and mechanical sensors for the monitoring of thermal and mechanical parameters. The potential impact of FOSs is related to their good metrological properties, their small size and their flexibility, as well as to their immunity from electromagnetic field. Their main advantage is the possibility to use textile based on fiber optic in a magnetic resonance imaging environment, where standard electronic sensors cannot be employed. This last feature makes FOSs suitable for monitoring biological parameters (e.g., respiratory and heartbeat monitoring) during magnetic resonance procedures. Research interest in combining FOSs and textiles into a single structure to develop wearable sensors is rapidly growing. In this review we provide an overview of the state-of-the-art of textiles, which use FOSs for monitoring of mechanical parameters of physiological interest. In particular we briefly describe the working principle of FOSs employed in this field and their relevant advantages and disadvantages. Also reviewed are their applications for the monitoring of mechanical parameters of physiological interest.

KEYWORDS

smart textiles; fiber optic sensors; fiber Bragg grating sensors; respiratory monitoring; macrobending sensors; hetero-core fiber optics; Magnetic Resonance Imaging; MR-compatibility

1. INTRODUCTION

The growing interest in smart textiles for medical applications is driven by the aim to increase the mobility of patients who need a continuous monitoring of physiological parameters [1,2,3,4]. Smart textiles are able to interact with the environment; therefore they embed one or more sensors to monitor various mechanical, thermal and chemical parameters (e.g., strain, temperature, displacement, oxygen blood saturation) [5,6,7]. During the last decades the use of fiber optic-based sensors (FOSs) has been gaining acceptance in a large number of applications in the fields of civil engineering, the automotive industry and medicine [8,9,10], among others. These sensors allow the measurement of physical and chemical parameters employing a large number of working principles and configurations [4]. FOSs can be divided into intrinsic sensors, where the fiber optic constitutes the sensing element, and extrinsic ones, where the fiber optic is only used as a medium to transport light. There are a number of reviews and books that focus on the description of FOSs

[11,12,13,14,15], hence it is not possible to describe all the applications and working principles of FOSs in one journal article.

FOSs have good metrological properties (i.e., low zero drift and sensitivity drift, good accuracy and good sensitivity and large bandwidth), they offer the possibility to implement distributed sensors and they are immune to electromagnetic interferences. These features make FOSs an emerging solution for the monitoring of physiological parameters and more generally for applications in medicine [9]. In fact, thanks to the mentioned valuable characteristics, FOSs can compete with other traditional electrical and mechanical sensors, and in many fields of application the superiority of their performances has been demonstrated. Lastly, the possibility to develop Magnetic Resonance (MR)-compatible sensors further motivates the growing interest in this technology.

Fiber optic technology is particularly attractive for application in smart textiles because it allows both sensing and signal transmission. Moreover, polymer optical fibers (POFs) are cheap, lightweight, flexible and robust, and they are able to measure high strain values without damage.

During the last decade, several groups of research have focused their efforts on obtaining a substantial development in the integration of smart textiles and fiber optic technology [16,17] as reported in a recent review [18].

In this paper, we aim at describing the use of smart textiles in medicine employing FOSs; in particular we focus on their use for monitoring physiological parameters by measuring mechanical variables and on the metrological properties and performances of the devices reported in literature. Also reviewed are the working principles of FOSs most frequently used in smart textiles for the mentioned applications. In Section 2 the working principles of the FOSs used in smart textiles and their applications in medicine are described. In Section 3 the applications in physiological monitoring of smart textiles based on FOSs along with their main advantages and drawbacks are reviewed.

2. WORKING PRINCIPLE OF FIBER OPTIC SENSORS USED IN SMART TEXTILES

FOSs can be designed using a large number of working principles. In this section the ones used to develop smart textiles for measuring mechanical variables in physiological monitoring are reviewed. In particular we focus on sensors based on the fiber Bragg grating technology and on intensity-based FOSs. The next subsections describe the working principle of these sensors with a brief description of their application in medicine.

2.1. Fiber Bragg Grating Sensors

Fiber Bragg grating (FBG) sensors consist of a periodic perturbation of the refractive index along the fiber core length obtained by exposure of the core to an intense optical interference pattern. Hill reported the first fabrication of FBGs in the late 1970s [19], but a major breakthrough occurred a decade later, when Meltz et al. improved the technique for their fabrication [20]. This last study can be considered the milestone, which enabled the development of FBG sensors for a large number of applications.

Basically, an FBG can be considered as a short segment of a fiber optic (usually FBG longer than 3 mm–6 mm are employed, although for particular application smaller FBG are required), which reflects a narrow range of wavelengths and transmits all others. A schematic representation of the working principle of an FBG sensor and of its response to strain is shown in Figure 1.

The wavelength of the input light that is back-reflected (λ_B) is sensitive to temperature and strain. In fact, λ_B can be expressed by the following equation, which is the first-order Bragg condition of the grating:

$$\lambda_B = 2 \cdot \Lambda \cdot \eta_{eff} \tag{1}$$

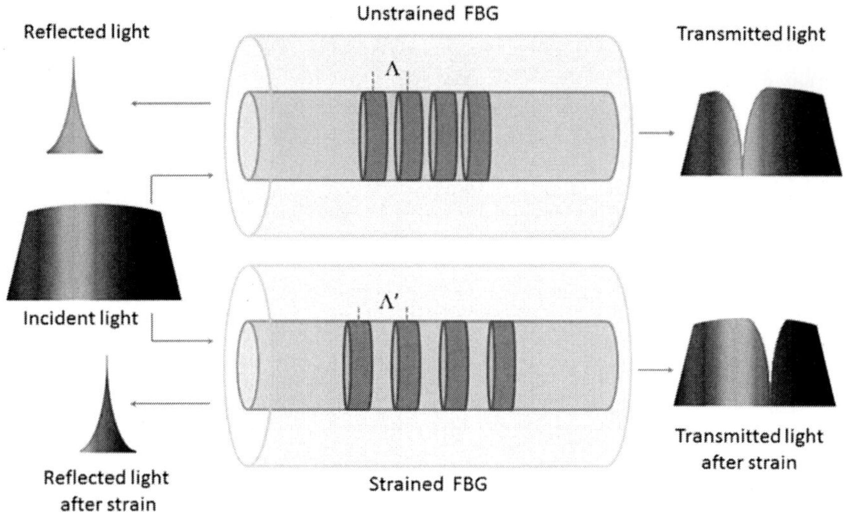

Figure 1. Schematic of the working principle of Fiber Bragg grating (FBG) sensors, and its response to strain.

where Λ is the spatial period of the grating and η_{eff} is the effective refractive index of the fiber core. Both the terms are sensitive to strain and temperature, as a consequence, the use of a proper configuration allows estimating strain and temperature or both, by monitoring changes of λ_B. The dependence of λ_B can be described by the fractional changes of Λ and η_{eff}:

$$\frac{\Delta\lambda_B}{\lambda_B} = \frac{\Delta\Lambda}{\Lambda} + \frac{\Delta\eta_{eff}}{\eta_{eff}} \tag{2}$$

Actually, due to the different sensitivity of Λ and η_{eff} to temperature changes and strain, the FBGs sensors used to estimate strain are based on the λ_B shift due to Λ changes, the FBG sensors used to monitor temperature are based on the λ_B shift due to η_{eff} changes. Equation (1) shows an important advantage of the FBG: their output is not affected by fluctuation of source intensity.

FBG sensors are employed in a large number of industrial fields to monitor different physical variables, including, among others, temperature, pressure, flow and vibrations. Moreover, FBGs have been largely used to monitor physiological parameters and more in general in medical fields, such as stroke volume, blood pressure and heartbeats [21,22], in microsurgery [23,24], foot pressure in diabetic patients [25], in biomechanical studies [26,27], in the monitoring of temperature during thermal ablation of cancer [28,29,30,31], in respiratory monitoring system [32,33], and in tactile sensing [34,35].

FBGs sensors are considered to hold great potential for application in medicine due to their good metrological properties. Moreover their characteristics of biocompatibility, non-toxicity and chemical inertness, as well as their small size and flexibility make them particularly attractive for invasive measurements during in vivo trials [36]. Lastly, they are suitable for application in environments with high electromagnetic noise, thanks to their immunity to electromagnetic interferences and their intrinsic MR-compatibility [37].

2.2. Intensity Modulated FOSs

Intensity-modulated FOSs modulate light intensity, measured by a secondary element (e.g., a photodiode), in response to an environmental effect. A simple configuration of this kind of sensor is shown in Figure 2A. Two optical fibers are held in close proximity to each other; the light is injected into one of the optical fibers; as the light expands into a cone of light, its intensity, emitted by the first fiber and conveyed into the other one, depends on the distance (d) between the two fiber tips. Therefore, the light intensity can be considered an indirect measurement of the distance between the two fibers and of other physical variables influencing this distance. A similar configuration can be designed either by using a single fiber and a mirror (Figure 2B) or by using more than one fiber to obtain a differential configuration. The differential solution allows neglecting the influence of the light source intensity on sensor output. Another configuration to develop intensity-modulated FOSs is underpinned by the phenomenon that the light is lost from an optical fiber when it is bent. In particular a bent radius causes a leakage of the light traveling within the core of the fiber into the cladding with a resulting intensity modulation of light propagating through an optical fiber (Figure 2C). Macrobending sensors based on hetero-core fibers have been proposed to measure several physical properties [38,39].

Intensity-modulated FOSs have been used in different medical fields starting from the late 1960s, when Lekholm and Lindstrom proposed a sensor for intravascular pressure monitoring [40]. A similar sensor was proposed to monitor the intracranial pressure [41], and commercially available intracranial pressure intensity-modulated FOSs are produced by Camino Laboratories Inc. Their performances are largely investigated in clinical settings [42]. Moreover these sensors are used for pressure and temperature monitoring [43].

Macrobending FOSs find applications in medicine mainly in the monitoring of respiratory movements [44,45,46]. These sensors are largely used in smart textiles, therefore their medical applications will be described in more detail in the following section. A particular approach based on the bending of the fiber is also used to develop flow sensors for mechanical ventilation [47,48].

3. SMART TEXTILES BASED ON FIBER OPTIC SENSORS: MEDICAL APPLICATIONS

Smart technical textiles are by definition textiles that can interact with their environment. Their ability to sense physical and chemical parameters can be accessed by embedding several kinds of sensors. The use of FOSs on textile is particularly attractive in some medical fields because of the possibility to use this technology during MR procedures, the cost reduction of key optical components and the improvement of the component quality, as well as the good metrological properties of these sensors. In particular, polymer optical fibers (POFs) match well with the requirements for application in smart textiles, being cost effective, lightweight and robust; moreover POFs are able to measure high strain values of several ten percent without fiber damage [49].

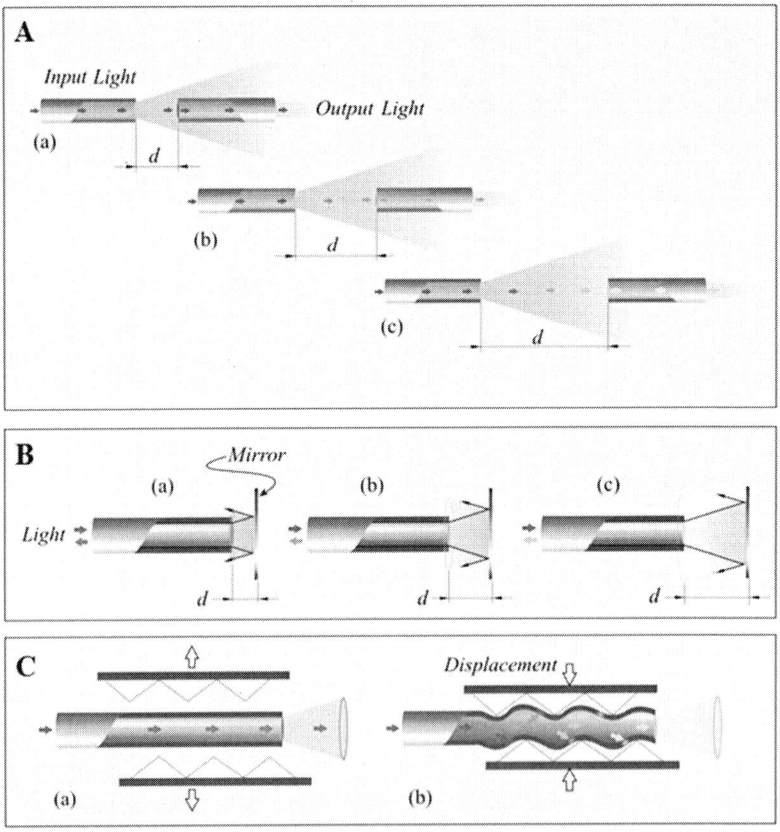

Figure 2. (**A**) Schematic of the working principle of an intensity modulated sensor using two fiber optic; (**B**) schematic of the working principle of an intensity modulated sensor using a fiber optic and a mirror; (**C**) schematic of the light lost from the fiber core caused by bending (adapted from [10]).

Significant developments of the integration of FOSs in textiles were driven by several groups of research, and in particular by the group involved in the European project OFSETH (optical fiber sensors embedded into technical textile for healthcare). OFSETH aimed at investigating the possibility to use FOSs embedded into textiles to monitor several physiological parameters.

In the following two subsections the application of smart textiles based on FBG sensors and on intensity modulated FOSs are described.

3.1. Smart Textiles Based on FBG Sensors: Medical Applications

FBG is one of the most frequently employed technologies to design smart textiles based on fiber optics, thanks to their good sensitivity to strain. This characteristic allows developing several configurations based on the measurement of the strain experienced by the FBG sensor to monitor different parameters of physiological interest.

In particular this solution has been employed to monitor respiratory movements. The use of FBG to monitor respiratory movements and breathing rate has been demonstrated in the past [32,50,51], but only during the last decade have they been embedded in smart textiles. A number of studies regarding this topic have been proposed by the groups involved in the OFSETH project. A smart textile embedding two different FOSs (i.e., FBG and macrobending) for respiratory monitoring has been developed. FOSs and the use of MR-compatible connectors allow the use of the proposed smart textiles on anesthetized patients during MR procedures. Indeed, these sensors are free from metallic or electrical conductive wires (when using a custom made MR-compatible connector); in addition, they are remotely interrogated via an optical fiber cable allowing the location of the monitoring unit outside of the MR field [52,53], as schematically reported in Figure 3.

The high FBG sensitivity to strain (i.e., ≈1.2 pm·$\mu\varepsilon^{-1}$) allows discriminating small strains; therefore they have been used to monitor thoracic movements that are smaller than abdominal ones. Two different methods to embed the FBG sensors within textiles (stitching and crochet) have been proposed. The calibration of the system, which embeds the FBG by stitching (see Figure 4A), has been performed by stretching the textiles in steps of 0.4% up to 40%. During the calibration, the FBG experiences strains up to 0.8%. The system shows good linearity from about 8.5% to 35%–40% of textile stretching, with a sensitivity of 0.35 nm/% and an accuracy better than 0.1% of elongation. Only preliminary experiments on the textiles developed with the crochet method have been performed. The integration of this sensor with a different FOS has been assessed on ten healthy volunteers [54]. Trials to evaluate the long-term properties and the stability of the respiratory sensor by an ad hoc developed simulator have been performed as well [55]. In particular the mechanical robustness of the sensor has been tested with more than 90,000 cycles in 129 hours with a simulated breathing rate between 10 breaths per minute and 12 breaths per minute [56]. A further valuable characteristic of the proposed system (see Figure 4B) is that it enables the continuous measurement of respiratory movement providing free access to all vital organs for medical staff actions [57].

A recent interesting study proposed a simple wearable system based on a single FBG sensor, which allows the simultaneous detection of both heart rate and respiratory cycles [58]. The main innovation is related to the structure in which the FBG sensor is embedded, which is [9] a PVC laminate resulting in a strain-sensitive foil, manufactured by an industrial spread-coating process. This integrated solution shows a sensitivity of 0.8 pm·$\mu\varepsilon^{-1}$. The authors performed trials on healthy volunteers, using two filters for breathing rate monitoring (band-pass filter tuned in the range 0.1–0.4 Hz) and cardiac frequency monitoring (band-pass in the range 0.5–1.3 Hz). The same group of research developed a system for breathing rate and cardiac frequency monitoring able to work in a wide range of temperature [59]. The authors selected the polychloroethanediyl as carrier material due to its excellent performance/cost and to the good sensitivity to strain. The system showed a linear response for elongation ranging from 0.6% to 1.6%. In this range the FBG output experienced an increase of about 8 nm; therefore the sensitivity was about 8 nm/%. They also investigated the sensors' output changes with changes in temperature and found an increase of Bragg wavelength of about 1.5 nm for a temperature increase of 140 °C (a thermal sensitivity of about 10.7×10^{-3} nm/°C).

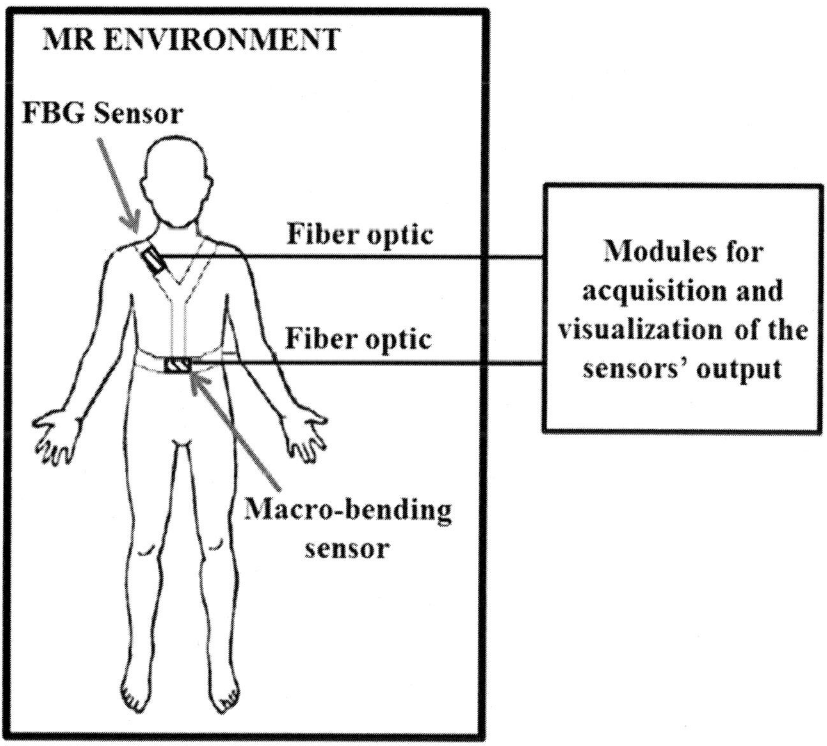

Figure 3. Schematic representation of the monitoring systems proposed by and developed in the optical fiber sensors embedded into technical textile for healthcare (OFSETH) project.

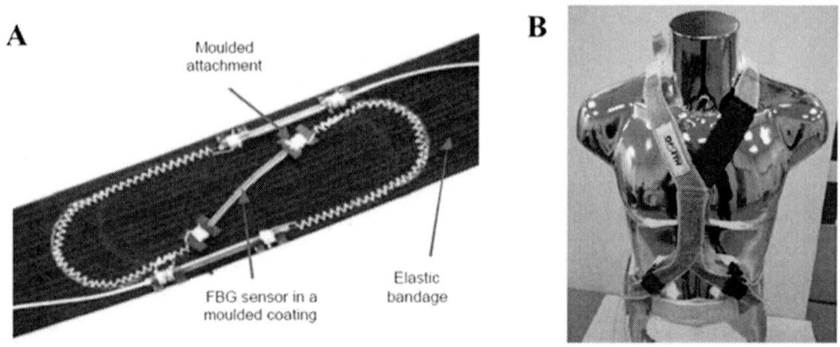

Figure 4. (**A**) Design of the FBG sensor developed in OFSETH project (adapted from [57]); (**B**) MRI-compatible sensing harness which embeds the fiber optic sensors for respiratory monitoring (adapted from [57]).

Other simple solutions have been proposed to monitor respiration and heart activity. In particular, Dziuda and coauthors developed a system which consisted of a Polymethyl methacrylate (PMMA) board with the size of 220 × 95 mm^2 and a thickness of 1.5 mm. An FBG sensor was attached with epoxy adhesive [60]. They experimentally assessed the error of the system in the measurement of breaths per minute and heartbeats per minute during magnetic resonance imaging examinations, showing promising results (about one breath per minute and about three heartbeats per minute). The same group of researchers reported a system based on two FBG sensors positioned orthogonally to each other [61]. This group validated their system on three patients during MRI procedures by comparing their results with the ones obtained by an MRI-compatible portable module [62]. The results were promising (a relative error lower than 8% can be considered satisfactory considering that the system is intended for monitoring rather than diagnosis). Recently, they proposed a system for heart rate monitoring. It is MR-compatible and has been tested on seven volunteers showing a root mean square error of less than six beats per minute [63].

Allsop and coauthors developed a wearable system for respiratory function monitoring [64]. Their system employs an array of 40 in-line FBG sensors that produce 20 curvature sensors at different locations, each sensor consisting of two FBGs. They carried out experiments to measure the absolute volumetric changes of the human torso and estimated an error of 6% on the average volume.

Lastly, Li and coworkers developed a wearable sensor in intelligent clothing based on FBG for human body temperature monitoring [65]. They partly embedded an FBG in a polymer filled strip to improve the sensitivity of the measuring system. This way they obtained a temperature sensitivity of 150 pm·°C^{-1}, almost 15 times higher than that of the bare FBG. To measure human body temperature, they distributed five FBGs in five places (i.e., left chest, right chest, left armpit, right armpit and at center of the upper back). Moreover, they developed a model to estimate the body temperature by the data of the five FBGs. With this method they found an accuracy of ±0.1 °C.

The main characteristics of the abovementioned smart textiles and their applications are reported in Table 1.

3.2. Smart Textiles Based on Intensity-Modulated FOSs

FOSs based on intensity modulation and in particular macrobending sensors are employed to develop smart textiles for monitoring of physiological parameters.

Textiles based on these kinds of FOSs were proposed by the groups of research involved in the OFSETH project. In particular, they developed smart textiles based on macrobending FOS for the monitoring of respiratory rate on patients during MRI procedures [52,53]. The sensors are MR-compatible and the monitoring devices are located out of the magnetic resonance environment (see Figure 3). A standard single mode fiber was embedded within a textile as shown in Figure 5A. This macrobending sensor, due to the lower sensitivity than the FBG sensors, was used to monitor abdominal movements, because these movements are larger than thoracic ones (Figure 5B). The authors found large oscillations in the sensor's output during the stretching of the textile, which could cause a wrong computation of the breathing rate. As a consequence, they designed a sensor with more periods of the U-shape on the textile bandage (Figure 5B). This solution allowed to substantially increase the sensitivity (e.g., for a textile elongation of 38%, the sensor response increased from less than 3 dB with a single-loop design to more than 28 dB with a 10 loops-design). Lastly they developed a sensor based on the optical time-domain reflectometry (OTDR) in polymer optical fiber (POF). Basically, the macrobending entails a change of the backscattering in POF that can be sensed by the technique of OTDR [56]. This solution allows measuring strain in different locations of a single fiber (distributed measurement). A respiratory simulator was employed to test the robustness of both the macrobending sensor and the sensor based on the OTDR technique. Both these sensors show low variations after cycles at 10 breaths per minute and elongation up to 3% and 5% for the OTDR sensor and for the macrobending one, respectively. These sensors were also tested on healthy volunteers.

An interesting solution to develop an MR-compatible sensor wearable sensor for respiratory monitoring based on intensity modulation was presented in [66]. It consists of a PMMA tube, a mirror, a spring and a plastic optical fiber. Abdominal movement causes a variation of the distance between the mirror and the distal end of the plastic optical fiber that is related to the intensity of reflected light coupled to the fiber (as shown in the schematic representation in Figure 2B). The authors also tested the sensor during an MR procedure, and they did not find any negative effects related to patient safety and image quality.

Table 1. Smart textiles and wearable systems based on fiber optic sensors: working principle, medical application and metrological properties.

Reference	Working Principle	Medical Application	Metrological Properties and Other Features
[52,53,56]	Silica FBGs	Respiratory monitoring during MRI procedures	Non-invasive; MR compatible; good linearity in a wide range of strains with sensitivity = 0.35 nm/%; accuracy better than 0.1% of elongation
[58,59]	Silica FBGs	Cardiac and Respiratory monitoring during MRI procedures	Non-invasive; MR compatible; sensitivity of 0.8 nm/$\mu\varepsilon^{-1}$
[59]	Silica FBGs	Cardiac and Respiratory monitoring	Non-invasive; Sensitivity of 8 nm/%; good linear trend; thermal sensitivity ≈ 10.7×10^{-3} nm/°C
[60,62]	Silica FBGs	Cardiac and Respiratory monitoring during MRI procedures	Non-invasive; MR compatible; Simple design; Good accuracy in terms of breathing rate (±1 bpm) and heartbeat per minute (±3 bpm); relative error in patients during MRI procedures <8%
[61]	Silica FBGs	Cardiac and Respiratory monitoring during MRI	Non-invasive; Simple design; Flat frequency response in the range of interest (0.5 Hz up to 20 Hz); maximum relative error of 12%
[63]	Silica FBG	Heart rate monitoring	Non-invasive; MR compatible; Root mean square error lower than 6 beats per minute
[64]	Silica FBGs	Respiratory function monitoring	Non-invasive; 6% of error on the average volume
[65]	Bare FBG	Body temperature monitoring	Non-invasive; Sensitivity of 150 pm/°C in the range of interest (from 33 °C to 42 °C); accuracy 0.1 °C
[52,53,56]	Macro-bending/OTDR technique	Respiratory monitoring	Non-invasive; MR compatible; Good sensitivity stability after 172800 cycles (variations < 10%); low cost component for their interrogation
[66]	Intensity modulated	Respiratory monitoring	Non-invasive; MR compatible; low cost component
[67,68]	Intensity modulated	Respiratory monitoring	Non-invasive; low cost component
[69,70]	Macrobending hetero-core fiber optic	Respiratory monitoring	Non-invasive; low cost component; good agreement with the breathing rate measured by a commercial device
[71]	microbending	Respiratory monitoring during MRI procedures	Non-invasive; MR compatible; Accuracy better than ±2 breaths per minute
[72]	microbending	Respiratory rate and heart rate	Non-invasive; MR compatible; Accuracy better than ±2 breaths or beats per minute for respiratory monitoring heart rate
[73]	microbending	Heartbeat and respiratory monitoring	Non-invasive; low cost component; good agreement with the heart beat measured by a commercial device

Krehel and coauthors developed a textile for respiratory monitoring based on FOS previously described [67,68]. Basically, the working principle of the sensor employed can be explained as follows: the fiber optic geometry changes when a force (or pressure) is applied on the fiber; these geometry changes affect the wave guiding properties and hence induce light loss in the optical fiber. They characterized the sensor in [67], showing a range of measurement for force applied on 3 cm of fiber length up to 40 N, with a discrimination threshold of 0.05 N. They also performed the feasibility assessment of the wearable system for breathing rate monitoring [68]. The trials were performed at two breathing rates and at different positions of the sensing textile on the human body. Lastly, the comparison with a commercial device for respiratory measurements was performed using the Bland Altman analysis. The results showed that a large part of the differences between the measurements obtained by the FOS textile and the commercial device was concentrated in the range ±3 min^{-1}, and the limits of agreement were about ±6 min^{-1}.

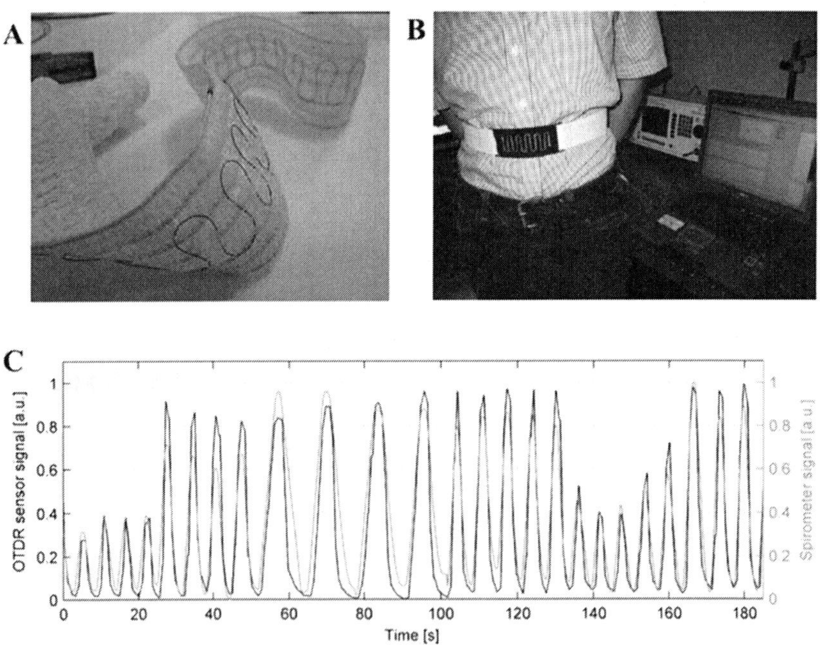

Figure 5. (**A**) Textiles based on macrobending FOS embedding a standard single mode fiber (adapted from [57]); (**B**) Macrobending sensors for monitoring of abdominal movements (adapted from [57]); (**C**) Output of the OTDR sensors during the monitoring of abdominal movement (adapted from [57]).

Alemdar and coauthors developed a smart textile based on macrobending hetero-core fiber optic for respiratory movement analysis [69,70]. They embedded within a textile, a periodic macrobending hetero-core fiber optic that formed a periodic “U”. They experimentally assessed the increase of sensitivity to strain with the number of loops (ranging from one to seven loops) and the influence of the loop length on sensitivity. Then, the most sensitive configuration was tested to monitor the abdominal movement on a healthy volunteer.

Lau et al. designed a simple microbend fiber optic sensor for respiratory monitoring during MRI procedures [71]. They assessed the sensor feasibility during MRI procedures on twenty healthy volunteers, showing an accuracy of ±2 breaths per minute in the measurement of frequency rate. They also developed a system to monitor both respiratory rate and heart rate with a similar approach [72]. They tested the system on 11 volunteers during an MRI procedure, showing an accuracy of two breaths per minute for respiratory rate monitoring, and two beats per minute for heart rate monitoring.

Recently, a smart textile based on the use of a periodic fiber optic microbend sensor has been proposed for heartbeat and respiratory monitoring [73]. The authors integrated a section of multimode optical fiber sandwiched between parallel strips acting as a microbender onto an elastic substrate. Basically, the system monitors respiration and heartbeat by detecting the vibration caused by these actions. The microbending results in light loss, which is detected by a photodetector. After a set of in vitro experiments, they tested the textile on healthy volunteers. The measured heartbeat was compared to the one measured by a commercial device, the measured respiratory rate with the number of cycles manually counted in one minute.

The main characteristics of the abovementioned smart textiles and their applications are reported in Table 1.

4. DISCUSSION

Smart textiles are used in a number of industrial fields such as civil engineering, transport and medicine, amongst others. Regarding the application in medicine and healthcare, they allow the continuous monitoring of physiological parameters of great importance (e.g., respiratory activity and heartbeat).

The monitoring of these parameters can be performed by several approaches, and different devices are commercially available. The main advantage to integrate FOSs into smart textiles is the possibility to develop MR-compatible systems. The number of MR scanners, the request for high field devices and for MR procedures in several medical branches (e.g., cardiology, surgery, orthopedics and neurology) are increasing worldwide. This trend reflects the growing need for MR-compatible systems able to monitor physical parameters inside the scanner, in order to provide real-time feedback about the status of the patient.

As a consequence, smart textiles based on the use of FOSs can be considered a potential new market niche in the field of healthcare monitoring. In this scenario is important to have a small size and lightweight system, and more in general a system allowing a normal range of motion and preserve the patient's comfort. As a consequence, intensity-based fiber optic sensors seem to be indicated more than FBG ones that require the use of the interrogator. Moreover, also in homecare application the privacy and confidentiality of the patient must be respected. Solutions, such as enclosed rooms without traffic or others present and data transmission over secure lines can help to fulfill this requirement. Lastly, the problems of washability and, in some cases, sterilizability of FOS-based smart textiles are not well addressed. This feature must be considered because it can be a cost-effective option and can avoid the use of these systems for mono-patient or disposable applications. Therefore, future studies should address the robustness of these textiles against stress and washing cycles or their sterilizability The advantage is that FOSs are intrinsically safer than conventional sensors as they do not have any electrical connection with the patient and may be used at medium-term for monitoring different physiological parameters in order to replace standard sensors during MRI procedures. In this scenario it is also important to take into account the issue related to the connection of fibers. As reported by Kinet et al. [74] different solutions can be employed (e.g., commercially available standard connectors for optical fibers and ad hoc designed connectors). For the applications in the MRI environment, it is crucial to employ ad hoc designed, MR-compatible connectors.

Moreover, they can answer to the increasing demand for wearable systems for continuous monitoring of physiological parameters. A comfortable and wearable measuring system may significantly improve the quality of life of patients.

The main concerns related to the extensive use of smart textiles based on FOSs and their spread in commercially available products are related to the necessity to improve their stability due to light coupling (in particular in homecare long-term applications), and the abovementioned issue related to the connection of fibers (in particular in application during MRI procedures). The continuous improvement of the quality of the key optical components and their cost reduction, as well as the good metrological properties of these sensors motivates the increasing interest in this topic.

A recent review published by Quandt et al. in 2014 [18], described body-monitoring systems based on optical fibers and their application in the field of physiological monitoring. In this review, we devoted our attention especially to smart textiles based on FOSs for monitoring mechanical parameters of physiological interest. Also briefly reviewed are the working principles of the FOSs most used in these applications. Lastly, we focused on the metrological properties and on the performances of the systems reported in literature. In particular the most employed FOSs are based on FBG technology and on the modulation of intensity (in particular on macrobending). They are largely used for heartbeat and respiratory monitoring. These two kinds of sensors show complementary advantages and disadvantages in terms of sensitivity to movement and costs. FBG sensors are able to measure small strains thanks to their high sensitivity; on the other hand, in order to have good performances, interrogation must take place through an expensive device. The interrogation of intensity based FOSs is very simple and involves low cost and compact components, and their integration into textile fabrics may be very straightforward; on the other hand they are less sensitive than FBG sensors and are usually employed to monitor large movements (e.g., abdominal movements).

The use of smart textiles based on FOS in respiratory and heartbeat monitoring is just in its beginning stages, but their good metrological performances and the possibility to monitor patients in real time during MRI procedures motivates to continue research effort devoted to this topic.

AUTHOR CONTRIBUTIONS

Carlo Massaroni, Paola Saccomandi and Emiliano Schena contributed equally to this manuscript; Carlo Massaroni, Paola Saccomandi and Emiliano Schena wrote the article.

REFERENCES

1. Lymberis, A.; Olsson, S. Intelligent Biomedical Clothing for Personal Health and Disease Management: State of the Art and Future Vision. Telemed. J. E-Health 2003, 9, 379–386.
2. Lobodzinski, S.S.; Laks, M.M. New devices for very long-term ECG monitoring. Cardiol. J. 2012, 19, 210–214.

3. Lobodzinski, S.S. ECG patch monitors for assessment of cardiac rhythm abnormalities. Prog. Cardiovas. Dis. 2013, 56, 224–229.

4. Inan, O.T.; Etemadi, M.; Wiard, R.M.; Giovangrandi, L.; Kovacs, G.T. Robust ballistocardiogram acquisition for home monitoring. Physiol. Meas. 2009, 30, 169–185.

5. Zhang, X.; Tao, X. Smart textiles: Passive smart. Text. Asia 2001, 45–49.

6. Zhang, X.; Tao, X. Smart textiles: Active smart. Text. Asia 2001, 49–52.

7. Krehel, M.; Wolf, M.; Boesel, L.F.; Rossi, R.M.; Bona, G.L.; Scherer, L.G. Development of a luminous textile for reflective pulse oximetry measurements. Biomed. Opt. Express 2014, 5, 2537–2547.

8. Li, H.N.; Li, D.S.; Song, G.B. Recent applications of fiber optic sensors to health monitoring in civil engineering. Eng. Struct. 2014, 26, 1647–1657. [CrossRef]

9. Da Silva, A.F.; Gonçalves, A.F.; de Almeida Ferreira, L.A.; Araújo, F.M.M.; Mendes, P.M.; Correia, J.H. PVC smart sensing foil for advanced strain measurements. IEEE Sens. J. 2010, 10, 1149 1155.

10. Silvestri, S.; Schena, E. Optical-fiber measurement systems for medical applications. In Optelectronics: Devices and Applications; Predeep, P., Ed.; InTech: Rijeka, Croatia, 2011.

11. Lee, B. Review of the present status of optical fiber sensors. Opt. Fiber Technol. 2003, 9, 57–79. [CrossRef]

12. Marazuela, M.; Moreno-Bondi, M. Fiber-optic biosensors—An overview. Anal. Bioanal. Chem. 2002, 372, 664–682.

13. Grattan, K.T.V.; Meggitt, B.T. Optical Fiber Sensor Technology, Vol. 2—Devices and Technology; Chapman & Hall: London, USA, 1998.

14. Udd, E. Fiber Optic Sensors, 1st ed.; John Wiley & Sons, Inc.: Hoboken, New Jersey, 2006.

15. Wang, X.D.; Wolfbeis, O.S. Fiber-optic chemical sensors and biosensors (2008–2012). Anal. Chem. 2012, 85, 487–508.

16. Krebber, K.; Liehr, S.; Witt, J. Smart technical textiles based on fibre optic sensors. In Proceedings of the 22nd International Conference on Optical Fiber Sensors, Beijing, China, 14 October 2012.

17. Ghosh, S.; Amidei, C.; Furrow, K. Development of a sensor-embedded flexible textile structure for apparel or large area applications. Indian J. Fibre Text. 2005, 30, 42–48.

18. Quandt, B.M.; Scherer, L.J.; Boesel, L.F.; Wolf, M.; Bona, G.L.; Rossi, R.M. Body-monitoring and health supervision by means of optical fiber-based sensing systems in medical textiles. Adv. Healthc. Mater. 2015, 4, 330–355.

19. Hill, K.O.; Fujii, Y.; Johnson, D.C.; Kawasaki, B.S. Photosensitivity in optical fiber waveguides: Application to reflection filter fabrication. Appl. Phys. Lett. 1978, 32, 647–649. [CrossRef]

20. Meltz, G.; Morey, W.W.; Glenn, W.H. Formation of Bragg gratings in optical fibers by a transverse holographic method. Opt. Lett. 1989, 14, 823–825.

21. Rao, Y.J.; Webb, D.J.; Jackson, D.A.; Zhang, L.; Bennion, I. Optical in-fiber bragg grating sensor systems for medical applications. J. Biomed. Opt. 1998, 3, 38–44.

22. Gurkan, D.; Starodubov, D.; Yuan, X. Monitoring of the heartbeat sounds using an optical fiber bragg grating sensor. In Proceedings of the 4th IEEE Conference Sensors 2005, Irvine, CA, USA, 30 October–3 November 2005.

23. Haines, D.E. Determinants of lesion size during radiofrequency catheter ablation: The role of electrode-tissue contact pressure and duration of energy delivery. J. Cariovasc. Electrophysiol. 1991, 2, 509–515. [CrossRef]

24. Yokoyama, K.; Nakagawa, H.; Shah, D.C.; Lambert, H.; Leo, G.; Aeby, N.; Ikeda, A.; Pitha, J.V.; Sharma, T.; Lazzara, R.; et al. Novel contact force sensor incorporated in irrigated radiofrequency ablation catheter predicts lesion size and incidence of steam pop and thrombus. Circ. Arrhythm. Electrophysiol. 2008, 1, 354–362.

25. Hao, J.Z.; Tan, K.M.; Tjin, S.C.; Liaw, C.Y.; Roy Chaudhuri, P.; Cuo, X.; Lu, Cu. Design of a foot-pressure monitoring transducer for diabetic patients based on FBG sensors. In Proceedings of the LEOS, the 16th Annual Meeting of the IEEE, Tucson, AZ, USA, 27–30 October 2003; pp. 23–24.

26. Al-Fakih, E.; Abu Osman, N.A.; Mahamd Adikan, F.R. The use of fiber Bragg grating sensors in biomechanics and rehabilitation applications: The state-of-the-art and ongoing research topics. Sensors 2012, 12, 12890–12926.

27. Tjin, S.C.; Tan, Y.K.; Yow, M.; Lam, Y.Z.; Hao, J. Recording compliance of dental splint use in obstructive sleep apnoea patients by force and temperature modeling. Med. Biol. Eng. Comput. 2001, 39, 182–184.

28. Saccomandi, P.; Schena, E.; Caponero, M.A.; di Matteo, F.M.; Martino, M.; Pandolfi, M.; Silvestri, S. Theoretical analysis and experimental evaluation of laser-induced interstitial

thermotherapy in ex vivo porcine pancreas. IEEE Trans. Biomed. Eng. 2012, 59, 2958–2964.

29. Di Matteo, F.; Martino, M.; Rea, R.; Pandolfi, M.; Panzera, F.; Stigliano, E.; Schena, E.; Saccomandi, P.; Silvestri, S.; Pacella, C.M.; et al. US-guided application of Nd: YAG laser in porcine pancreatic tissue: An ex vivo study and numerical simulation. Gastrointest. Endosc. 2013, 78, 750–755.

30. Saccomandi, P.; Schena, E.; Giurazza, F.; del Vescovo, R.; Caponero, M.A.; Mortato, L.; Panzera, F.; Cazzato, R.L.; Grasso, F.R.; di Matteo, F.M.; et al. Temperature monitoring and lesion volume estimation during double-applicator laser-induced thermotherapy in ex vivo swine pancreas: A preliminary study. Laser Med. Sci. 2014, 29, 607–614. [CrossRef]

31. Saccomandi, P.; Schena, E.; di Matteo, F.M.; Pandolfi, M.; Martino, M.; Rea, R.; Silvestri, S. Laser Interstitial Thermotherapy for pancreatic tumor ablation: Theoretical model and experimental validation. In Proceedings of the 2011 Annual International Conference of the IEEE on Engineering in Medicine and Biology Society, EMBC, Boston, MA, USA, 30 August–3 September 2011; pp. 5585–5588.

32. Wehrle, G.; Nohama, P.; Kalinowski, H.J.; Torres, P.I.; Valente, L.C.G. A fibre optic Bragg grating strain sensor for monitoring ventilatory movements. Meas. Sci. Technol. 2001, 12, 805. [CrossRef]

33. Allsop, T.; Miller, M.; Bennion, I.; Carroll, K.; Lloyd, G.; Webb, D.J. Application of long-period-grating sensors to respiratory plethysmography. J. Biomed. Opt. 2007, 12. [CrossRef]

34. Cowie, B.M.; Webb, D.J.; Tam, B.; Slack, P.; Brett, P.N. Fibre Bragg grating sensors for distributive tactile sensing. Meas. Sci. Technol. 2007, 18. [CrossRef]

35. Saccomandi, P.; Schena, E.; Oddo, C.M.; Zollo, L.; Silvestri, S.; Guglielmelli, E. Microfabricated Tactile Sensors for Biomedical Applications: A Review. Biosensors 2014, 4, 422–448.

36. Mishra, V.; Singh, N.; Tiwari, U.; Kapur, P. Fiber grating sensors in medicine: Current and emerging applications. Sens. Actuat A Phys. 2011, 167, 279–290. [CrossRef]

37. Taffoni, F.; Formica, D.; Saccomandi, P.; Pino, G.D.; Schena, E. Optical fiber-based MR-compatible sensors for medical applications: An overview. Sensors 2013, 13, 14105–14120.

38. Watanabe, K.; Tajima, K.; Kubota, Y. Macrobending characteristics of a hetero-core splice fiber optic sensor for displacement and liquid detection. IEICE Trans. Electron. 2000, 83, 309–314.

39. Nishiyama, M.; Sasaki, H.; Watanabe, K. A deformation sensitive pad-structure embedded with hetero-core optic fiber sensors. Sens. Actuat A Phys. 2007, 136, 205–211. [CrossRef]

40. Lekholm, A.; Lindstrom, L. Optoelectronic transducer for intravascular measurements of pressure variations. Med. Biol. Eng. 1969, 7, 333–335.

41. Wald, A.; Post, K.; Ransohoff, J.; Hass, W.; Epstein, F. A new technique for monitoring epidural intracranial pressure. Med. Instrum. 1977, 11, 352–354. [PubMed]

42. Piper, I.; Barnes, A.; Smith, D.; Dunn, L. The Camino intracranial pressure sensor: Is it optimal technology? An internal audit with a review of current intracranial pressure monitoring technologies. Neurosurgery 2001, 49, 1158–1165. [PubMed]

43. Wolthuis, R.A.; Mitchell, G.L.; Hartl, J.C.; Saaski, E. Development of a dual function sensor system for measuring pressure and temperature at the tip of a single optical fiber. IEEE Trans. Biomed. Eng. 1993, 40, 298–302.

44. Babchenko, A.; Khanokh, B.; Shomer, Y.; Nitzan, M. Fiber optic sensor for the measurement of respiratory chest circumference changes. J. Biomed. Opt. 1999, 4, 224–229.

45. Maletras, F.X.; Augousti, A.T.; Mason, J. Construction and calibration of a new design of Fiber Optic Respiratory Plethysmograph (FORP). In Proceedings of the SPIE 4444, Optomechanical Design and Engineering 2001, San Diego, CA, USA, 29 July 2001; p. 285.

46. Nishyama, M.; Miyamoto, M.; Watanabe, K. Respiration and body movement analysis during sleep in bed using hetero-core fiber optic pressure sensors without constraint to human activity. J. Biomed. Opt. 2011, 16. [CrossRef]

47. Schena, E.; Saccomandi, P.; Silvestri, S. A high sensitivity fiber optic macro-bend based gas flow rate transducer for low flow rates: Theory, working principle, and static calibration. Rev. Sci. Instrum. 2013, 84. [CrossRef]

48. Saccomandi, P.; Schena, E.; Silvestri, S. A novel target-type low pressure drop bidirectional optoelectronic air flow sensor for infant artificial ventilation: Measurement principle and static calibration. Rev. Sci. Instrum. 2011, 82. [CrossRef]

49. Husdi, I.R.; Nakamura, K.; Ueha, S. Sensing characteristics of plastic optical fibres measured by optical time-domain refelectometry. Meas. Sci. Technol. 2004, 15, 1553–1559. [CrossRef]

50. Augousti, A.T.; Maletras, F.X.; Mason, J. Improved fibre optic respiratory monitoring using a figure-of-eight coil. Physiol. Meas. 2005, 26, 585–590.

51. Augousti, A.T.; Raza, A.; Graves, M. Design and characterisation of a fibre optic respiratory plethysmograph (FORP). In Proceedings

of the SPIE 1996, Biomedical Sensing, Imaging, and Tracking Technologies I, San Jose, CA, USA, 27 January 1996; pp. 250–257.

52. Narbonneau, F.; Kinet, D.; Paquet, B.; Depré, A.; de Jonckheere, J.; Logier, R.; Zinke, J.; Witt, J.; Krebber, K. Smart textile embedding optical fibre sensors for healthcare monitoring during MRI. Adv. Sci. Tech. 2008, 60, 134–143. [CrossRef]

53. Grillet, A.; Kinet, D.; Witt, J.; Schukar, M.; Krebber, K.; Pirotte, F.; Depré, A. Optical fiber sensors embedded into medical textiles for healthcare monitoring. IEEE Sens. J. 2008, 8, 1215–1222. [CrossRef]

54. De Jonckheere, J.; Jeanne, M.; Narbonneau, F.; Witt, J.; Paquet, B.; Kinet, D.; Kreber, K.; Logier, R. OFSETH: A breathing motions monitoring system for patients under MRI. In Proceedings of the Annual International Conference of the IEEE Engineering in Medicine and Biology Society, Minneapolis, USA, 2-6 September 2009; pp. 1016–1019.

55. Narbonneau, F.; D'Angelo, L.T.; Witt, J.; Paquet, B.; Kinet, D.; Kreber, K.; Logier, R. FBG-based smart textiles for continuous monitoring of respiratory movements for healthcare applications. In Proceedings of the 12th IEEE International Conference on e-Health Networking Applications and Services (Healthcom), Lyon, France, 1–3 July 2010; pp. 277–282.

56. Witt, J.; Narbonneau, F.; Schukar, M.; Krebber, K.; de Jonckheere, J.; Jeanne, M.; Kinet, D.; Paquet, B.; Depré, A.; D'Angelo, L.T.; et al. Medical textiles with embedded fiber optic sensors for monitoring of respiratory movement. IEEE Sens. J. 2012, 12, 246–254. [CrossRef]

57. Krebber, K. Smart Technical Textiles Based on Fiber Optic Sensors. In Current Developments in Optical Fiber Technology; InTech: Rijeka, Croatia, 2013; pp. 319–344.

58. Silva, A.F.; Carmo, J.P.; Mendes, P.M.; Correia, J.H. Simultaneous cardiac and respiratory frequency measurement based on a single fiber Bragg grating sensor. Meas. Sci. Tech. 2011, 22. [CrossRef]

59. Carmo, J.P.; da Silva, A.M. F.; Rocha, R.P.; Correia, J.H. Application of fiber Bragg gratings to wearable garments. IEEE Sens. J. 2012, 12, 261–266. [CrossRef]

60. Dziuda, Ĺ.; Skibniewski, F.W.; Krej, M.; Baran, P.M. Fiber Bragg grating-based sensor for monitoring respiration and heart activity during magnetic resonance imaging examinations. J. Biomed. Opt. 2013, 18. [CrossRef]

61. Dziuda, Ĺ.; Skibniewski, F.W.; Krej, M.; Lewandowski, J. Monitoring respiration and cardiac activity using fiber Bragg grating-based sensor. IEEE Trans. Biomed. Eng. 2012, 59, 1934–1942.

62. Dziuda, L.; Krej, M.; Skibniewski, F.W. Fiber Bragg grating strain sensor incorporated to monitor patient vital signs during MRI. IEEE Sens. J. 2013, 13, 4986–4991. [CrossRef]

63. Krej, M.; Dziuda, L.; Skibniewski, F. A method of detecting heartbeat locations in the ballistocardiographic signal from the fiber-optic vital signs sensor. IEEE J. Biomed. Health Inform. 2015. [CrossRef]

64. Allsop, T.; Bhamber, R.; Lloyd, G.; Miller, M.R.; Dixon, A.; Webb, D.; Castañón, J.D. A.; Bennion, I. Respiratory function monitoring using a real-time three-dimensional fiber-optic

shaping sensing scheme based upon fiber Bragg gratings. J. Biomed. Opt. 2012, 17. [CrossRef]

65. Li, H.; Yang, H.; Li, E.; Liu, Z.; Wei, K. Wearable sensors in intelligent clothing for measuring human body temperature based on optical fiber Bragg grating. Opt. Express 2012, 20, 11740–11752.

66. Yoo, W.J.; Jang, K.W.; Seo, J.K.; Heo, J.Y.; Moon, J.S.; Park, J.Y.; Lee, B.S. Development of respiration sensors using plastic optical fiber for respiratory monitoring inside MRI system. J. Opt. Soc. Korea 2010, 14, 235–239. [CrossRef]

67. Krehel, M.; Rossi, R.M.; Bona, G.L.; Scherer, L.J. Characterization of flexible copolymer optical fibers for force sensing applications. Sensors 2013, 13, 11956–11968.

68. Krehel, M.; Schmid, M.; Rossi, R.M.; Boesel, L.F.; Bona, G.L.; Scherer, L.J. An optical fibre-based sensor for respiratory monitoring. Sensors 2014, 14, 13088–13101.

69. Alemdar, K.; Likoglu, S.; Fidanboylu, K.; Toker, O. A novel periodic macrobending hetero-core fiber optic sensor embedded in textiles. In Proceedings of the SPIE 9062, Smart Sensor Phenomena, Technology, Networks, and Systems Integration, San Diego, CA, USA, 8 March 2014.

70. Alemdar, K.; Likoglu, S.; Fidanboylu, K.; Toker, O. A novel periodic macrobending hetero-core fiber optic sensor embedded in textiles. In Proceedings of the 8th International Conference on Electrical and Electronics Engineering (ELECO), Bursa, Turkey, 28–30 November 2013; pp. 467–471.

71. Lau, D.; Chen, Z.; Teo, J.T.; Ng, S.H.; Rumpel, H.; Lian, Y.; Yang, H.; Kei, P.L. Intensity-Modulated Microbend Fiber Optic Sensor

for Respiratory Monitoring and Gating During MRI. IEEE Trans. Biomed. Eng. 2013, 60, 2655–2662.

72. Chen, Z.; Lau, D.; Teo, J.T.; Ng, S.H.; Yang, X.; Kei, P.L. Simultaneous measurement of breathing rate and heart rate using a microbend multimode fiber optic sensor. J. Biomed. Opt. 2014, 19.

73. Yang, X.; Chen, Z.; Chia, S.; Lam, H.; Ng, S.; Teo, J.; Wu, R. Textile Fiber Optic Micro Bend Sensor Used for Heartbeat and Respiration Monitoring. IEEE Sens. J. 2015, 15, 757–761. [CrossRef]

74. Kinet, D.; Mégret, P.; Goossen, K.W.; Qiu, L.; Heider, D.; Caucheteur, C. Fiber Bragg Grating Sensors toward Structural Health Monitoring in Composite Materials: Challenges and Solutions. Sensors 2014, 14, 7394–7419.

Index

H

I

K

M

O

P

R

S

T

W